Locomotives and Rail Road Transportation

Avinash Kumar Agarwal · Atul Dhar
Anirudh Gautam · Ashok Pandey
Editors

Locomotives and Rail Road Transportation

Technology, Challenges and Prospects

Springer

Editors
Avinash Kumar Agarwal
Department of Mechanical Engineering
Indian Institute of Technology Kanpur
Kanpur, Uttar Pradesh
India

Anirudh Gautam
Rolling Stock Division
RITES Ltd.
Lucknow, Uttar Pradesh
India

Atul Dhar
School of Engineering
Indian Institute of Technology Mandi
Kamand, Himachal Pradesh
India

Ashok Pandey
Center of Innovative and Applied Bioprocessing (CIAB)
Mohali, Punjab
India

ISBN 978-981-10-3787-0 ISBN 978-981-10-3788-7 (eBook)
DOI 10.1007/978-981-10-3788-7

Library of Congress Control Number: 2017930134

Printed on acid-free paper

This Springer imprint is published by Springer Nature
The registered company is Springer Nature Singapore Pte Ltd.
The registered company address is: 152 Beach Road, #22-06/08 Gateway East, Singapore 189721, Singapore

Preface

Rail transport has played a very significant role in development of modern civilization over past 150 years. Due to its capability of bulk transportation of passengers and goods, it has proved to be more environmentally benign and energy efficient on a per capita basis, compared to any other mode of transportation such as road or air. Due to large system size and longer service life of involved systems, technological advancements in the field of combustion engine technology, alternative fuels, emission control technologies, system design and control, etc., are being implemented at a comparatively slower pace in railways across the globe. However, due to increased global awareness of global warming and air pollution, railways are committing themselves for setting up and meeting more stringent emission norms worldwide.

An international workshop, 3rd ISEES Workshop on "Sustainable Energy, Environment & Safety with Railway Centric Theme", was held at Research Designs and Standards Organisation (RDSO), Lucknow, India during December 21–23, 2015 under the aegis of International Society for Energy, Environment and Sustainability (ISEES). This workshop provided a platform for discussions between eminent scientists and engineers from various countries including India, USA, South Korea, Thailand and Austria. In this workshop, eminent speakers presented their views related to different aspects of technology developments related to railroad transportation, use of numerical tools and modeling tools, and use of sophisticated experimental techniques, which enhanced our understanding of locomotive combustion technology. In addition, there is a great deal of interest in emissions control and use of advanced materials, production and utilization methods of various alternative fuels in conventional IC engine-based power trains, sophisticated and reliable control of big and complex energy systems in railroad transportation sector.

In recent past, lot of developmental activities related to reduction of emissions, using exhaust heat recovery system for electrical power generation, turbocharging, space-heating, increasing mechanical output and other feasible applications for increasing overall power train efficiency have been undertaken. For planning strategic implementation of these advancements in railways, an integrated and comprehensive plan needs to be developed. This research monograph is an effort in this direction and contains the main topics covered in the workshop and provides the latest developments in this domain. Main theme of this monograph is technological development of locomotive for overcoming current challenges related to energy saving, emission reduction and improving passenger comfort. Various chapters focus on effective utilization of fuels, production and utilization of other non-conventional fuels, emissions and noise reduction, effective power utilization and power production from waste heat, and fundamental study of combustion processes for increasing efficiency and reducing emissions.

The editors would like to express their sincere gratitude to the authors for submitting their work in a timely manner and revising it appropriately at a short notice. We would like express our special thanks to Dr. Dhiraj V. Patil (IIT Mandi), Dr. P. Anil Kishan (IIT Mandi), Dr. Srikrishna Sahu (IIT Madras), Dr. Krithika Narayanaswamy (IIT Madras), Sh. Sunil Patahk (IIP Deharadun), Prof. Tarun Gupta (IIT Kanpur), Dr. Rakesh Kumar Maurya (IIT Ropar), Dr. Dhananjay Kumar Srivastava (IIT Kharagpur) and Akhilendra Pratap Singh (IIT Kanpur), who reviewed various chapters of this monograph and provided their valuable suggestions to improve the manuscripts. We acknowledge the support received from various funding agencies and organizations for the successful conduct of the ISEES workshop, where these monographs germinated. These include Department of Science and Technology, Government of India (Special thanks to Dr. Sanjay Bajpai); RITES Ltd., India (Special thanks to Sh. Pradeep Gupta); Office of Naval Research Global, Singapore (Special thanks to Dr. Ramesh Kolar); TSI, India (Special thanks to Dr. Deepak Sharma); Caterpiller India; AVL India; Dynomerk Controls, India (Special thanks to Sh. Kishore Raut); CEI Softwares, India; ESi Group, Pune; BHEL India; and Bosch India.

We hope that researchers in various fields related to locomotive technologies such as emission control, efficiency improvement, alternative fuel production and utilization, noise and vibrations control will find this monograph helpful.

Kanpur, India — Avinash Kumar Agarwal
Kamand, India — Atul Dhar
Lucknow, India — Anirudh Gautam
Mohali, India — Ashok Pandey

Contents

About the Editors

Prof. Avinash Kumar Agarwal joined IIT Kanpur in 2001. Professor Agarwal was at ERC, University of Wisconsin, Madison, USA as a Postdoctoral Fellow (1999–2001). His areas of interest are IC engines, combustion, alternative fuels, hydrogen, conventional fuels, lubricating oil tribology, optical diagnostics, laser ignition, HCCI, emissions and particulate control and large bore engines. Professor Agarwal has published more than 200 peer-reviewed international journal and conference papers. He is Associate Editor of ASME Journal of Energy Resources Technology and International Journal of Vehicle Systems Modelling and Testing. He has edited "Handbook of Combustion" (5 Volumes; 3168 pages), published by Wiley VCH, Germany. Professor Agarwal is a Fellow of SAE (2012), Fellow of ASME (2013) and a Fellow of INAE (2015). Professor Agarwal is the recipient of Prestigious Shanti Swarup Bhatnagar Prize-2016 in Engineering Sciences.

Dr. Atul Dhar is Assistant Professor at IIT Mandi since 2013. He received his M.Tech. and Ph.D. degrees from Department of Mechanical Engineering, IIT Kanpur in 2006 and 2013, respectively. Atul Dhar graduated from HBTI, Kanpur in 2004. He was awarded the Erasmus Mundus fellowship of European Union for pursuing postdoctoral research at Ecole Centrale de Nantes, France in 2013. He is recipient of young scientist award from ISEES in 2015. His areas of interest include reciprocating IC engines, emission control, alternative fuels and lubricating oil tribology. He has co-authored more than 30 international peer-reviewed journal papers.

Dr. Anirudh Gautam works with the Indian Railway Service of Mechanical Engineers. He holds degrees in Mechanical and Electrical engineering from SCRA scheme of Jamalpur, a Master's in Quality Management from BITS Pilani, and a Master's in Engineering in Engine Design from University of Wisconsin, Madison. He has served in various roles in the Indian Railways. He later worked at Diesel Locomotive Works including engine manufacturing, engine design and transfer of technology. He built India's first 4000 hp diesel locomotive indigenously under ToT with EMD, General Motors, USA, for which he was awarded a national award by the Railway Minister. He has successfully completed various R&D projects at Engine Development Directorate, RDSO. He completed his Ph.D. from IIT Kanpur in 2013. He has two patents, 10 technical papers and 50 technical reports to his credit. He is presently on deputation to RITES as Group General Manager Rolling Stock Design.

Prof. Ashok Pandey is Eminent Scientist at the Center of Innovative and Applied Bioprocessing, Mohali. His major research interests are in the areas of microbial, enzyme and is Eminent Scientist at the Center of Innovative and Applied Bioprocessing, Mohali. His major research interests are in the areas of microbial, enzyme and bioprocess technology. He has to his credit over 1150 publications/communications, including 16 patents, 50+ books, 140 book chapters, 423 original and review papers, an *h* index of 78 and ~25,000 Goggle Scholar citations. Professor Pandey is the recipient of several fellowships such as the Fellowships of Royal Society of Biology (UK), Academician of European Academy of Sciences and Arts, Germany; ISEES; National Academy of Science (India); BRSI; and awards such as Thomson Scientific India Citation Laureate Award, USA; UNESCO Professor; Raman Research Fellowship Award, CSIR; and GBF, Germany. Professor Pandey is Editor-in-chief of *Bioresource Technology*, Honorary Executive Advisor of *Journal of Water Sustainability* and *Journal of Energy and Environmental Sustainability*, Subject editor of *Proceedings of National Academy of Sciences (India)* and serves on the editorial board of several other journals.

Part I
General

Introduction to the Locomotives and Rail Road Transportation

Avinash Kumar Agarwal, Atul Dhar, Anirudh Gautam and Ashok Pandey

Abstract Development of locomotive and advancement of rail road transportation is important for reducing emissions and becoming less dependent on conventional fossil fuels. Utilization of available alternative fuels such as methanol, DME and biodiesel can resolve energy crises in the foreseeable future. Effective use of exhaust heat recovery can be helpful in increasing overall efficiency and power generation. After-treatment devices are now a necessity to meet the upcoming emission norms for the rail road sector. For meeting the challenges of energy and environment, there is a need for advanced technological development in locomotive and rail road transportation sector.

Keywords EHR system · Exhaust gas after-treatment · Greenhouse gas (GHG) emissions · Noise and vibrations · Biofuels · DME

In terms of per capita goods transport and passenger travel, railways are certainly most energy efficient, fast and environment friendly transportation mode. However in the developing countries, technological advancement in rail road sector is lagging behind the technological advancement in road transport sector. Absence of emission norms for railway locomotives in most countries of the world is a typical fact highlighting this lack of technological advancement in the rail road transport sector.

A.K. Agarwal (✉)
Department of Mechanical Engineering, Indian Institute of Technology Kanpur, Kanpur 208016, Uttar Pradesh, India
e-mail: akag@iitk.ac.in

A. Dhar
School of Engineering, Indian Institute of Technology Mandi, Mandi 175005 Himachal Pradesh, India

A. Gautam
RITES Ltd., Lucknow 227208, Uttar Pradesh, India

A. Pandey
Center of Innovative and Applied Bioprocessing (CIAB), Mohali 160071 Punjab, India

A.K. Agarwal et al. (eds.), *Locomotives and Rail Road Transportation*,
DOI 10.1007/978-981-10-3788-7_1

Railways are also geared towards reducing their carbon footprint. For example Indian Railway (IR) has already started implementation of strategies to minimize environmental impact by developing carbon neutral infrastructure in new stations, which will reduce the carbon foot print for railways in long run [1]. IR is also working for improvement of energy efficiency, especially in the traction system, in order to reduce emissions.

Locomotives are large compression ignition engines using diesel. Large size of engine, long operating hours and older engine technologies lead to huge emissions of gases and particulate. The locomotive engines still use technology, which was used 30–40 years ago in automotive sector. Advancement in engine technology in diesel locomotives will be certainly helpful in reducing these emissions. Moreover absence of emission norms in rail road sector in many countries including India does not compel railways to use low emission technologies. In United States, vast reduction in CO_2 and NO_x emissions was achieved after implementing locomotive regulations [2]. There is 70% reduction in NO_x and particulate emission in locomotives meeting tier 4 standards, in comparison to locomotives meeting tier 3 standards and for this, mostly in-cylinder emission control approaches were used. Exhaust gas recirculation (EGR) is the most commonly used in-cylinder NO_x emission reduction technique. Implementation of properly designed EGR scheme can be helpful in simultaneous particulate and fuel consumption reduction or maintaining them in acceptable limits while reducing NO_x emissions significantly [2]. Particulate emissions in exhaust are solid particles consisting of condensed volatile matter around carbon core. They can also be reduced by improving in-cylinder fuel-air mixing and exhaust gas after-treatment. Detailed study of combustion processes reveals that evolution of soot consists of formation of nuclei, inception or nucleation and conglomeration, surface growth and oxidation of particles. Soot emissions can be reduced by efficient utilisation of fuel and homogeneous distribution of fuel in the combustion chamber. Study of soot formation with flame characteristics is a new area of research, which will have huge impact on soot reduction, especially in the context of large size locomotive engines. Different after-treatment devices are used for reduction of emission such as (i) diesel oxidation catalytic convertors (DOC), which reduces CO, HC and soluble organic fraction (SOF); (ii) diesel particulate filter (DPF), which traps soot from the engine emissions and (iii) selective catalytic reduction devices, which reduce NO_x emissions into N_2. All these have great potential for locomotive engines.

Environmental and public health aspects compel researcher to reduce noise levels from the locomotive engines. Main parameters responsible for locomotive noise are locomotive engine technology, use of noise suppression devices, track condition, track roughness, number of bridges and other track crossings, weight of coach, imperfect contact between coaches, length of the train, speed, vibrations and site area for propagation of noise waves. In various countries, different models are being used for numerical analysis of noise from locomotives. Passenger comfort and quality perception of the travellers is highly dependent on noise, vibrations, and harshness (NVH) characteristics of trains and locomotive is one of the main sources for these. Various devices and equipment are developed for finding solution to this problem. There is room for improvement in the NVH characteristics of locomotives and trains in rail road sector.

Significant amount of heat is wasted in locomotives via cooling system and exhaust gas. Part of this waste heat can be recovered. Energy recovery is important because this energy production involves use of conventional petroleum fuels. Diesel production from crude petroleum involves fuel extraction, fuel processing, and fuel delivery. In the locomotive, combustion of diesel converts chemical energy of fuel into thermal energy and finally it is converted into mechanical power which is then converted to loco motion. A sizeable amount of heat energy remains unused in this conversion process and is rejected into the atmosphere via exhaust system and cooling system. This wasted heat energy can be partially recovered with the help of thermoelectric generator (TEG) and organic Rankine cycle (ORC), which can be used for meeting auxiliary power requirements of the train and/or for increasing the mechanical power output i.e. traction capacity of the locomotive. A typical locomotive at heavy-haul loses 2 MW energy via the radiator. With 2% conversion efficiency of TEG, 50 kW energy can be recovered by the application of TEG in the radiator. TEG can recover up to 100 kW energy from the locomotive exhaust gas (at temperature 350 °C). Efficiency of ORC based turbine is much higher than TEG however they have installation and operational challenges [3].

Total dependence on conventional fuel (diesel) comes with threat of interruption in supply in future for rail road transportation sector. Therefore efforts are needed for migration towards alternative fuels, which have nearly identical combustion characteristics as that of diesel but with relatively lower emissions. These alternative fuels can be produced from locally available renewable resources. Biodiesel, alcohols (methanol and ethanol) and Di-methyl ether (DME) are some examples of available alternative fuels. Among various prevailing alternative fuel options for diesel engines, combustion related properties of biodiesel are closer to mineral diesel. Due to similar properties, it is accepted as a locomotive fuel based on the availability. Availability of methanol in bulk quantities and its clean burning characteristics due to absence of carbon-to-carbon bonds, makes it a suitable alternative fuel for locomotives. Methanol produces lower particulate and sulphur oxides and it has higher cetane rating than mineral diesel. However it has blending issues with mineral diesel and is poisonous in nature. Di-methyl ether (DME) is a clean burning alternative fuel, which has nearly same physical properties as that of liquefied petroleum gas (LPG). DME has unique auto ignition characteristics and very high cetane number. If the problems related to pressurised fuel injection system, fuel supply system, inferior lubricating properties and low calorific value can be resolved, DME can emerge as an excellent alternative fuel for locomotives with soot-less emission spectra.

Various issues, opportunities and challenges related to development of locomotive engine are covered in this monograph. Specific topics of this monograph include:

- Introduction to the Locomotives and Rail Road Transportation
- Diesel Locomotives of Indian Railways: A Technical History

- Exhaust Heat Recovery Options for Diesel Locomotives
- Diesel Locomotive Noise: Sources, Reduction Strategies, Methods and Standards
- Biodiesel as an Alternate Fuel for Diesel Traction on Indian Railways
- Fuel Properties and Emission Characteristics of Dimethyl Ether in a Diesel Engine
- Potential of DME and Methanol for Locomotive Traction in India: Opportunities, Technology Options and Challenges
- Exhaust After-treatment System for Diesel Locomotive Engines—A Review
- Catalytic Control Options for Diesel Particulate Emissions Including that from Locomotive Engines
- Soot Formation in Turbulent Diffusion Flames: Effect of Differential Diffusion
- Development of a Mobile Emission Test Car for Indian Railways

The above topics have been categorised into four groups: (i) General, (ii) Efficiency improvement and noise reduction, (iii) Alternate fuels for locomotive traction, (iv) Locomotive emission reduction and measurement.

The first section deals with technical history and general description of Indian railways. Indian railway is the second largest railways network in world, which is now planning for major technology improvements for improving the operational efficiency of railway infrastructure and improving the service quality. IR consumes approximately 2.6 billion liters of diesel per year [4]. ALCO-DLW diesel-electric engine, Electro-Motive diesel (EMD) engine, diesel multiple unit engines are main types of locomotives used in India. ALCO-DLW diesel-electric engine contains 6, 12 and 16 cylinder and each cylinder produces 200–225 hp. These cylinders are water cooled, turbocharged and after-cooled. However these engines do not meet any international emission standard. EMD engines have 16 cylinders and each cylinder produces 280 hp. These cylinders are water cooled, turbocharged and after-cooled. EMD engines also do not meet any international emission standard as of now. Diesel multiple units are equipped with Cummins KTL 50L engines, which meets the US EPA tier 0 standards [4].

The second section deals with efficiency improvement and noise reduction by continuous improvements in design and structure of locomotives. Description of NVH characteristics identifies various components of locomotive for reduction of noise and vibrations, describes measurement methods and suggests measures for reducing them in locomotives. Exhaust heat recovery is a promising method for improving the overall efficiency of locomotives. Larger size of diesel locomotives compared to the space available makes the installation of exhaust heat recovery systems in diesel locomotives more practical. In this chapter, feasibility and suitability of various exhaust heat recovery options for diesel locomotives has been discussed.

Third section on alternate fuels for diesel locomotive traction deals with use biodiesel, methanol and DME. Indian railways have already allowed 5% or higher biodiesel blends in diesel locomotive, depending on availability. Technical feasibility, opportunities, challenges and potential of DME and Methanol as locomotive fuels in India is also discussed. These fuels are ultra-low-emission fuels, which have

been overlooked in both energy policy and industry discussions, despite many attributes, which make them attractive for locomotives. DME is especially suitable in diesel locomotives due to its fuel characteristics. Motivation to use DME as a fuel candidate in locomotives comes from its soot-free emission due to absence of C–C bonds. It can be blended with mineral diesel to overcome limitations of using 100% DME namely low viscosity and density. Similarly Methanol can be blended with mineral diesel for use in locomotives however it has blend stability issues. Methanol has a high latent heat of vaporization; highest oxygen content amongst fuels, is sulfur free and has higher burning speed. When burned at high temperature, it can reduce smoke and NO_x emissions from CI engines.

Fourth section deals with locomotive emission measurement and reduction strategies for making them environment friendly. One chapter presents detailed review of after-treatment emission control options, which are relevant to locomotives. This chapter also covers fundamental investigations on soot-formation mechanism. Another chapter describes design, development and use of mobile emission measurement test car for locomotives for gaining necessary data for developing emission norms.

Technical aspects presented in this monograph include after-treatment options, energy recovery options, noise reduction options and approaches, alternative fuels for reducing carbon foot print, and combustion improvement for improving overall performance of locomotives for rail road transport sector. The information here is aimed to provide knowledge for development of various technologies, challenges and opportunities related to locomotives in terms of emission reduction and energy savings. The content is expected to give useful information to the end user, researchers and development engineers working in the field of rail road transport sector.

References

1. http://indianrailways.gov.in/railwayboard/upload/
2. Jeihouni Y, Franke M, Lierz M, Tomazic D, Heuser P Waste heat recovery for locomotive engines using the organic Rankine cycle. In: Proceedings of the ASME 2015 internal combustion engine division fall technical conference ICEF2015-1015, Nov 8–11, Houston, TX, USA
3. Francesco S, Juergen P (2010) Enhanced locomotive efficiency through waste heat recovery. In: Conference on railway engineering wellington, Sept 12–15
4. http://www.ecmaindia.in/Uploads/image/6imguf_Dr.AvinashAgarwalandDr.AnirudhGautam (IIT) Panel-4.pdf

Diesel Locomotives of Indian Railways: A Technical History

Joydeep Dutta and Avinash Kumar Agarwal

Abstract In this article, we would like to trace the history of the growth of diesel traction of Indian railways without completely sacrificing technical details. It starts from the very early times and discusses at length the coming of the ALCO locomotives on the Indian Railways (IR) and the reappearance of EMD locos on IR. We have tried to mix the historical facts with technical facts therefore calling the study presented in this paper as technical history.

1 The Early Diesels

Since this is a technical article on the diesel locomotives of the Indian Railways (IR), we urge the reader to first have a look at the appendix, where the locomotive codes are explained and then it would be easier to go through this article.

Diesel locomotives on IR have a long history. However the early history of diesel locomotion on IR is not very clear. It is usually assumed that the WDS1 broad-gauge diesel shunters worked in the Bandra area of Bombay (now Mumbai) in the 1930s. However Terry Martin [1] in his wonderful study on the Darjeeling Himalayan Railways, mentioned that one of the first successful dieselization in India occurred on Gaekwad's 2 ft 6 in. narrow-gauge Baroda State Railway in 1932, when Armstrong-Witworth sold four 80 hp diesel railcars. The same company supplied 90 hp diesel electric railcars to the Kalka-Shimla railway in 1933. In fact Terry Martin [1] wrote about an attempt to use diesel traction on the Darjeeling Himalayan Railways (DHR) when their directors put an order for a diesel locomotive fitted with a 165 hp General Motors (GM) diesel engine with Walford

J. Dutta (✉)
Economics Group, Department of Humanities and Social Sciences,
Indian Institute of Technology Kanpur, Kanpur 208016, India
e-mail: jdutta@iitk.ac.in

A.K. Agarwal
Engine Research Laboratory, Department of Mechanical Engineering,
Indian Institute of Technology Kanpur, Kanpur 208016, India

A.K. Agarwal et al. (eds.), *Locomotives and Rail Road Transportation*,
DOI 10.1007/978-981-10-3788-7_2

Transport Limited, which had its main office in Calcutta (now Kolkata). The unit cost was 23,950 rupees. However this locomotive was not successful on the DHR. Terry Martin [1] also claimed that this was one of the first diesel locomotives to be built in India. This had a mechanical drive with a five speed gear box.

Though IR now appears to be a broad gauge railway, the first mainline dieselization on IR began with the introduction of YDM1 locomotives in 1955 on the meter-gauge and they were first homed in Gandhidham on the Western Railways. There were 20 such locomotives with hydraulic transmission. One of the first beneficiaries of dieselization in India was the Northeast Frontier Railway. Its important rail junction Siliguri was the first home to YDM3 class diesel-electric locomotives built by the Electromotive Division of the GM. These locomotives were transferred later on to the Abu Road diesel shed of the Western Railways. The GM was the ruling king among diesel locomotive manufacturers. IR had however decided that though they need to move from steam to diesel, they did not want to remain captive to the big international manufacturers. They wanted to get a technology blueprint of a simpler locomotive and manufacture it in India. Though they approached GM for a technology transfer, GM was reluctant. However they did supply India with the superb 2400 hp WDM4 class, Co-Co diesel-electric locomotives, which were an export version of SD-24 class locomotives used on the railroads in the USA. These were supplied in 1962 and homed in Mughalsarai with a specific shed built for them. These locomotives were praised by most railwaymen associated with diesel traction in India. Only 72 of them were bought. Meanwhile IR experimented with hundred WDM1 class diesel-electric locomotives manufactured by American Locomotive Company (ALCO) in 1958. Dubbed as the "World Locomotive" by ALCO, these locomotives were mainly used in the Eastern and South Eastern Railways. Fitted with an ALCO class 244, V-12 diesel engine producing 1800 hp, it was used mainly for coal trains, but was also used for dieselization of some important trains like the Howrah-Madras Mail. It might be interesting to speculate why ALCO christened the WDM1 as 'World Locomotive'. These locomotives were provided with swing bolster, double suspension "World" Co-Co bogies. The term "World" refers to the fact that WDM1, which was called DL-500-C could work across various railways of the world. This locomotive had a 16 tonne axle load thus it could work in places where high-axle loads were not possible but had demand for faster trains. Apart from the IR, DL-500-C had worked in many other railways, for example RENFE in Spain. Once inside the locomotive cab, one would be surprised that it still used a steam-era whistle chord. In fact it has one control stand and boasted a dead-man's paddle, which the driver had to keep his foot on while driving. The drivers however were clever enough to just put a brick on the dead-man's paddle to save them from the irritation to keeping one of their feet on the paddle throughout. The WDM1 had a good visibility and was of the carbody design where all the equipment including the cab were hosed under a single casing.

IR at that period of time operated all trains with vacuum brakes, a legacy of the British Raj though air-brakes were more efficient. If one walks down from the cab of the WDM1 through the equipment section, one would first come across the electrical compartment, which consists of the main DC generator and an auxiliary generator and then the prime-mover, the V-12 ALCO 244 four-stroke diesel engine and then the compressor-exhauster unit. This exhauster was a special fitment for the IR, since the exhauster is used to create the vacuum in the train brake pipe. With a gear-pinion ratio of 92:19, this locomotive was capable of reaching upto 122 km/h though as per IR practice, it was restricted to 105 km/h. It had a fuel tank with a capacity of 3000 l diesel. At a speed of around 22 km/h, the maximum continuous tractive effort was achieved by the WDM1 which was about 19,280 kg. Continuous tractive effort is the tractive effort that can be maintained by the locomotive for a great length of time without damaging the traction motors. For more details on tractive effort, see the Wikipedia article (https://en.wikipedia.org/wiki/Tractive_force). Note that one should not confuse it with maximum tractive effort, which is achieved at much lower speeds but cannot be maintained for a long-time. The WDM1 locomotive did not have a dynamic brake, which was indeed a disadvantage for heavy freight operation and a more powerful and well equipped diesel locomotive was sought by the IR. The first ALCO shed was built in Gaya on the then Eastern Railway to home the WDM1 class locomotives. It was later dismantled, when the route was electrified.

Since higher horsepower locomotives were required by the IR, they originally approached the GM for a technology transfer, which as we mentioned, did not work out. However ALCO agreed to transfer the technology of DL560C class 2400 hp locomotive with a V16 engine, dubbed as ALCO 251B. In fact the ALCO 251-B prime mover was delivering 2600 hp under ideal test conditions, which translated to 2400 hp braking horsepower, that too under ideal test conditions, as prescribed by the Association of American Railroads (AAR). It was rechristened as the WDM2 and was capable of operating at a maximum speed of 120 km/h on IR and it became the most reliable broad gauge (5 ft 6 in. gauge) locomotive in the history of IR (Fig. 1). Though it was bought in as an intermediate step before electric traction takes over, however to the dismay of many electrical engineers on IR, the WDM2 had a complete sway over the IR operations. It was a locomotive adapted to operate in our dusty conditions and was very versatile in operating freights and cross-country mail/express trains. The testimony of the operating capability of DL 560C was borne out by the fact that it operated with an amazing reliability on the Andes mountain route to Cuzco from Lima, which was in 1960s, the highest altitude railroad in the world. The WDM2 was a Co-Co locomotive fitted with ALCO-trimount bogies of asymmetrical design with traction motors fitted in an asymmetrical fashion too. This bogie caused track damage on IR initially but then IR tweaked the design in a way that the tracks are not harmed. In fact this Co-Co trimount bogie did a lot of track damage on the US railroads, which finally led to ALCO going out of favour with important railroad companies and then closing down altogether in 1969. In fact, ALCO had licensed the technology to companies in developed countries like England, Spain, France, Australia and Canada. India was the only developing country included in this list by ALCO.

Compared to the WDM4, the machinery of the WDM2 was very simple and thus was easy to maintain by engaging a workforce, which had just moved into diesel from steam. The only new thing was the electrical components as the WDM2 was a diesel electric locomotive, for which the staff was required to be trained. The WDM4

Fig. 1 A WDM2 class locomotive built by DLW, homed at Pune Shed of Central Railway

needed specially trained staff. In fact ALCO completely transferred the technology of WDM2 to IR and helped it build its first indigenous diesel locomotive manufacturing facility in Varanasi, which was named as the Diesel Locomotive Works (DLW). It first began by assembling the knocked down kits sent from ALCO. This first such assembled kit was the WDM2 numbered 18233, which was dedicated to the service of the nation by the then Prime Minister Shri Lal Bahadur Shastri. ALCO meanwhile supplied around 121 fully assembled WDM2 to IR. DLW later on built WDM2's on its own and carried out several design modifications. More than 50 years of its coming to IR, the higher horsepower variants of the WDM2 family still continue to serve the IR and the nation. For more details on the WDM2, and various ALCO locomotives of IR, we would request the reader to consult the only book written about Indian ALCOs, which is titled *"The Story of Indian ALCOs: Legend of the WDM2"* by S. M. Sharma and J. Dutta [2].

It will indeed be a good idea to take a closer look at the WDM2 locomotive. The WDM2 is a road-switcher design, if we use the terminology of the US Railroads. It has a short-hood and a long-hood housing, the machinery with a cab in between the long and short-hoods with an inspection walkaway. We would take a walk from the short-hood peep into the cab and then walk along the inspection walkways and look at various machines in the WDM2. The short-hood houses the dynamic brake grids and the blower motor to cool the grids. It also houses the braking system. In fact, the early WDM2 were equipped only with vacuum brakes and used a Westinghouse braking system called 28LV1. When DLW started making WDM2's capable of both, air-brake and vacuum-brake operation, the braking system was called 28-LAV-1. The battery knife-edge switch, which is a prime requirement before starting a dead locomotive, is also housed in the short-hood. Now once inside the cab, one would find WDM2 to be quite ergonomic by the standards of the 1960s. There are two control stands. One for the short-hood operations and the other one for the long-hood operations. In both cases, the driver sits on the right, while the assistant sits on the left. The visibility of the WDM2 in both operations was pretty good and the drivers were very happy with these sturdy machines. Originally backrests were provided in the driver's seat however they were removed later on. We say that the layout was ergonomic in the sense that the control stand was clearly divided into two parts. The pneumatic part consisted of the gauges for the braking system like brake cylinder and brake pipe pressure, auxiliary reservoir and main reservoir pressure and a vacuum level measuring gauge along with the train brake handle (A9) and independent locomotive brake handle (SA9). Among electrical parts, we have the throttle, which can be operated over a range of eight notches. It has a reverser handle and also has a dynamic braking handle, an ampere meter, an electrical speedometer (on the long-hood control stand), warning lights, head lights and classification light switches, multiple-unit shut down switch and so on. If we sit in the short-hood control stand facing the short-hood, the wall separating the cab and the long-hood houses the electrical contactor, which are used to supply the current to the DC traction motors on the axles. It also houses the engine control switch and the traction motor cut-out switch. Then as we leave the cab and move down the walk way, we first find the electrical cabinet having the main generator and the auxiliary generator. Just on the inspection door is the front-truck traction motor blower. As we open the inspection door, we see huge ALCO 251-C prime mover and its Woodward governor. Further down, we have the compressor-exhauster cabinet followed by the radiator room with a huge radiator fan and a water tank attached to the roof, from where the water is supplied across the main engine to keep it cool. Inside the radiator compartment, apart from the huge radiator fan, one can see the encasing of lube-oil filters. The WDM2 has a big fuel tank (5000 litres capacity) and two main air reservoirs hung on two different sides. The one on the right side is for braking and the other one on the left side is for other pneumatic activities such as sander, windshield wiper, etc. For more details, please refer to ref [2]. With a gear-pinion ratio of 65:18, ALCO DL 560-C gave the maximum speed of 129 km/h with 17 MT axle load. On IR, WDM2 had 18.8 MT axle load and with the same gear-pinion ratio, its maximum speed was restricted to 120 km/h.

The WDM2 locomotive came in several series of road numbers. They were in 18 series, 17 series and 16 series and built in that order. The first WDM2 i.e. the class leader was 18040, which is now on display in the National rail Museum in Delhi. However the first WDM2 to reach the Indian shores was 18046. In fact from 18,000 to 18039, IR had numbered the WDM4 s the great rivals for WDM2. The WDM2's made by ALCO achieved under test conditions a continuous tractive effort of 28580 kg at a speed of 18 km/h.

The WDM2's have a life of 36 years. In fact one of the most important features of the approach to dieselization on IR was the standardization of WDM2 as the main broad gauge locomotive. Hence all over India, it was the same locomotive and a particular locomotive could be attended to, even if it was hundreds of kilometres away from its home shed. This had not been the characteristic of development of electric traction in India, which experimented with various classes of locomotives, making the maintenance of an electric locomotive very difficult and challenging, if it was far from the home shed or home railways. This simple standardization of diesel locomotives turned the WDM2 into the most reliable locomotive in the entire history of the IR. We must also take a careful note that all future developments of diesel locomotives done by the Indian railways were based on ALCO technology on WDM2 platform.

Looking at the success of the WDM2, IR also went in for an ALCO locomotive for the meter-gauge sections. This locomotive classified DL 535 by ALCO was rechristened as YDM4 in India (Fig. 2). Several locomotives of this class were built by the Montreal Locomotive Works (MLW), Canada, which was an ALCO subsidiary. Unlike the broad-gauge designs, the YDM4 was fitted with an ALCO 6-cylinder in-line 251-D engine, which gave 1350 hp and a brake horsepower in the range of 1100–1200 hp. Without any competition from electrics, this locomotive

Fig. 2 A DLW manufactured YDM4 meter-gauge diesel of the ALCO design

continued to hold its dominance over the vast meter-gauge network in India. These locomotives were capable of going to a maximum speed of 96–100 km/h and thus allowed introduction of many crack-trains on some important meter-gauge routes. The DLW started building this locomotive from 1968 till early 90s before the project-unigauge slowly took many YDM4's out of service. Many of them still work in Malaysia. In fact, the locomotives built in MLW were rechristened YDM4A and were used for hauling crack meter-gauge expresses like the Delhi-Jaipur Pink City Express and the Madurai-Chennai Egmore Vaigai express. These trains had very tight schedules, which would be envied even by the broad-gauge Shatabdi express trains. The YDM4 was a road-switcher like the WDM2 but had only one control stand. So if the loco ran with the short-hood face, then the driver sat on the right and while operating in the long-hood mode the driver sat on the left. In fact the Engine control switch (ECS) is in the control stand unlike the WDM2, which was in the electrical cabinet. The YDM4 had fuel tank quite similar in looks to that of WDM1 and had a capacity of 3000 l. With a gear-pinion ratio of 92:19, the locomotive ran at a maximum speed of 96 km/h, which was tweaked to 100 km/h on the YDM4A's, which led IR to introduce many meter gauge crack trains.

A very interesting experiment was carried out on the south central railways around 1970s, where they bought around eight diesel-hydraulic locomotives, which operated at a speed of 120 km/h. Two of them were fitted with Mercedes-Benz engines, while others were using a transmission designed by an Indian engineer M. M. Suri. All the eight locomotives were built by Henschel, Germany. Though they were very sophisticated locomotives (Christened as WDM3) and were efficient, their maintenance cost increased since they could not operate in dusty environment, typical to India. The only resource for WDM3 seems to be the website of the Indian Railways Fan Club (http://www.irfca.org/faq/faq-loco2d.html).

On the narrow gauge however, the diesel-hydraulic locomotives became the mainstay with ZDM3 and ZDM4 still operating Kalka-Shimla railway and NDM6 on the Darjeeling Himalayan Railway. These locomotives were in fact built at the Chittaranjan Locomotive Works (CLW), Kolkata, which is mainly an electric traction building unit but has developed facilities to build diesel hydraulic locomotives. They built famous shunters of WDS4 class, which were seen in major terminus stations and later replaced by DLW built WDS6 class 1350 hp diesel-electric shunters, having a YDM4 power-pack under a broad-gauge hood.

2 Improving WDM2

It was realized during 1980s that in order to keep up with the growing demand of railroad transportation in India, a higher-horsepower version of WDM2 locomotive was essential. The Engine Development Directorate of the Research Design and Standards Organizations (RDSO), Lucknow has been making efforts for quite some time since 1980s to improve the horsepower of ALCO 251-B prime mover. They could successfully uprate it to produce 3100 hp over a period of time. This could be done by design and development of double helix fuel injection pump, use of steel cap pistons and optimised turbochargers, in addition to few other changes. In fact at the Charbagh diesel shed, Lucknow, a WDM2 locomotive (18589) was fitted with this higher horsepower engine, which is now called 251-C. Instead of the standard DC/DC transmission, they put in an AC/DC transmission bought from General Electric (GE), Canada. They also fitted it with a higher capacity fuel tank. Under test conditions, the locomotive produced 2800 hp, which was a significant jump over the original WDM2. Thus was born the WDM2C which was later rechristened as WDM3A (Fig. 3) (do not confuse with WDM3, which was diesel hydraulic locomotive). This locomotive thus became the benchmark for development of higher horsepower diesels in India. Further it was realised that time has come to give-up ALCO Co-Co trimount trucks and develop separate locomotives for freight and passenger services. The WDG2C, which was finally called WDG3A with the 251-C, 3100 hp prime mover and high-adhesion bogies, became an instant success in freight operations (Fig. 6). IR wanted to move forward and have higher horsepower locomotives. The latest in that line was WDM3D, which was to have a 3400 hp power-pack however in actual operating conditions, it produced only 3300 hp. The WDM-3D was a micro-processor controlled locomotive, which performed in a reliable manner. There was also a model called WDP3A, which had a potential to operate up to 160 km/h and was of a very unique car body design, in sharp contrast to the traditional hood design of IR. This locomotive worked on crack express trains in the northern part of the country well into the mid-2000s and then was shifted to

Fig. 3 A WDM2 rebuilt as 3100 hp WDM3A by DMW patiala

ordinary passenger services. This class of locomotives still haul the Rajdhani Express from Hazrat Nizamuddin to Trivandrum Central in the Baroda-Trivandrum segment of the journey. An earlier experience in building the passenger locomotive was gained by using an upgraded version of ALCO-244 V-12 engine, which is now rated at 2300 hp. These locomotives with Bo-Bo bogies and classified as WDP1 were thought to be good for fast intercity services however they didn't perform well in practice.

The Diesel Modernization Work (DMW) at Patiala, formerly known at the Diesel Component Works (DCW) regularly carries out mid-life rehabilitation of WDM2 class locomotives and rebuilds them as WDM3A class locomotives. See [2] for more details Fig. 4.

IR had very recently (2013) developed a very unique diesel locomotive, where the diesel engine was replaced by three 800 hp diesel generator sets. This is a very energy efficient locomotive and looks almost like the ones found on US railroads. One can use all three, only two or just one generator set as per the duty assigned to the locomotive. It has been named WDM2G, and is made for light passenger trains. In fact, the current developed by the generator sets is fed to the traction motors and has a capacity of running at 120 km/h. Only two have been built until now and homed in Itarasi.

Before we end this section, it would be interesting to mention the role which IIT Kanpur has played in the development of the first Electronic Fuel Injection (EFI) system for the ALCO 251-C, 3100 hp diesel prime mover. This was achieved by a collaboration of Engine Research Laboratory (ERL, www.iitk.ac.in/erl), IIT Kanpur under the stewardship of Prof. Avinash Kumar Agarwal and Engine Development Directorate (EDD), RDSO under the stewardship of Dr. Anirudh Gautam.

Fig. 4 WDG3A class locomotives with high-adhesion bogies for heavy freight operations

Fig. 5 A long-hood view of the EFI locomotive (Photo courtesy: Prof. Avinash K. Agarwal)

Fig. 6 A short-hood view of the EFI locomotive (Photo courtesy: Prof. Avinash K. Agarwal)

An old WDM2 (16502) was rebuilt at DMW, Patiala as a 3100 hp WDM3A (16502R) in 2011 (Figs. 5, 6), which was retrofitted with the EFI system developed at ERL, IIT Kanpur. This was the first locomotive based on the ALCO platform to have an EFI system anywhere in the world. This locomotive was unveiled in August 2011 at DMW, Patiala and then homed at the Charbagh Diesel Shed, Lucknow. This locomotive showed superior engine performance in comparison with the standard WDM3A. An electronic fuel injection system for a 4-stroke, 16 cylinders, V-configuration, medium speed, large bore locomotive diesel engine has been developed and successfully retrofitted on a rebuilt diesel locomotive. The engine employed a Pump-Line-Nozzle (PLN) system for fuel injection into the cylinder. Original fuel injection system used was a mechanical fuel injection pump connected to a mechanical fuel injector through a high pressure fuel line. The fuel injection pump metered the fuel delivery using a single helix machined on its plunger. The fuel injection timings were however optimized only for the rated speed and load resulting in non-optimised operation at other engine operating points. An electronic fuel injection pump having a solenoid valve for both fuel metering and injection timing along with ECU were developed for retrofitment on rebuilt diesel locomotives. Interfacing of the ECU to the engine test cell controller was done by developing suitable hardware and software. ECU calibration was done and various maps of the engine were developed. The engine was tested on the engine test bed at EDD, RDSO, Lucknow. High pressure injector, modified fuel headers, fuel connection systems, a new high capacity fuel pump and layout of the wire harness were installed. After thorough testing and debugging, the EFI kit was retrofitted on a rebuilt diesel locomotive at DMW Patiala and tested on load box followed by brief field trials. EFI system delivered 3.3% fuel saving in passenger duty cycle and 3.97% in freight duty cycle. In addition there was an appreciable reduction in the smoke emissions during steady-state as well as transient operations. For more details see [3].

3 Back to the GM-EMD

The GM diesels had been on the IR's mind and they also knew that big manufacturer will allow technology transfer, if they are on a back foot in their domestic market. Finally GM-EMD was on the back foot in the US market, while facing a stiff competition from GE. To stay on in the market, EMD introduced the first diesel locomotive with 3-phase AC asynchronous traction motors. This locomotive classified as SD70MAC showed a great potential in hauling coal trains. IR in fact used this opportunity to transfer the technology of SD70MAC to India. GM agreed this time for the technology transfer in order to increase their global footprint and to stay afloat in the locomotive market. They then built the GT46MAC and GT46PAC, which were the export versions of SD70MAC with separate passenger and freight versions. These became the now famous WDG4 and WDP4 diesel locomotives of the IR. Fitted with EMD 710 V-16 two-stroke diesel engines producing 4000 hp, these locomotives with AC traction motors were a complete revolution in the diesel locomotive scenario in India. Having high adhesion bogies of the High Tension Steel Cast (HTSC) type, micro-processor control and an ergonomic cab, these locomotives immediately became a hallmark of advancement of diesel traction in India.

Let us now have a brief look into the way a WDP4 or a WDG4 functions. These locomotives have a diesel engine that runs a traction alternator (AC generator), which produces a single-phase AC, which gets rectified as DC, whose voltage can be varied as per the throttle position (Fig. 7). This is called the DC link voltage, which is fed into the two Traction Control Cabinets (TCC1 and TCC2), consisting of Traction Inverters and their associated computers. They covert this DC link voltage to variable frequency 3-phase AC voltage and then feed it into the traction motors. In fact the computers in TCC1 and TCC2 are linked to the main EM2000 computer in the cab, thus the power output is always known to the drivers. Further an important feature of the EMD locomotive is that when the traction motor is overloaded, the on-board computers can reduce the field excitation of the traction alternator irrespective of the throttle position. The traction inverters were originally provided by Siemens and were based on GTO technology. An Indian company 'Medha' provides the traction inverters based on IGBT technology, which allowed an increase of horsepower to 4500 hp. In fact in Siemens based TCC1 and TCC2 each controlled only one bogie however the indigenous IGBT technology allowed control of each axle individually. This is a very important improvement in the technology of EMD locos and in fact, the IGBT based locos are called GT46ACe.

Fig. 7 HTSC bogies on the WDG4 and dynamic braking grids and fans

Fig. 8 Radiator compartments on the WDP4 class EMD locomotives

The loco has one hood and the cab is at one end. Because of the design of the radiator, there were some visibility issues during long-hood operation thus IR indigenously developed a dual cab version called WDP4D and WDG4D (Fig. 8).

IR is keen on improvement in the horsepower output further. On a trial basis, they built a huge 20 cylinder 5500 hp locomotive called WDG5. These EMD locomotives are now changing the face of IR operations. GM however has sold EMD and EMD now calls itself Electromotive Diesels and changed several hands in short period of time.

Important Note: Many of the original data from ALCO was collected from the library of the National Rail Museum (NRM), Delhi, while the lead author was researching on the book [2]. There were several advertisements given by ALCO in the Indian Railways Magazine of 1963/1964 where they also provided detailed data. The lead author had collected the data from those sources. Authors would thus like to thank the National rail Museum, Delhi and the then Director of NRM, Mr. Mayank Tewari for helping and facilitating author's research. The lead author learnt about the important features of EMD locos in 2008 from DLW, Varanasi while researching on the book [2], a very little part of which has been mentioned here.

The lead author would like to note that it is very difficult to carry out academic research on the history of Indian diesel locomotives due to lack of books and academic papers. Most data are with the railways and few facts in this chapter have

Fig. 9 Front profile of the WDG4 class 4000 hp EMD freight locomotives

been obtained by discussion with various railway employees while working on the book [2] Fig. 9.

Acknowledgements We are greatly indebted to Mr. J. L. Singh for discussing the early history of the Indian diesel locomotives specially the WDS2, YDM1 and YDM3. Mr J. L. Singh was an officer in the Indian Railways. He left the railways in 1993, where his last posting was at the Indian Railway Institute for Mechanical and Electrical Engineers (IRIMEE) at Jamalpur. He taught the students about diesel locomotives. He retired finally from RITES (Rail Indian Technical and Economical services Limited) in 2005 as an Executive Director. All photos in this article except the two photographs of the EFI locomotive are taken by Aprurva Bahadur of Pune, who is an engineer by profession but is a leading expert on diesel locomotives and also one of India's foremost photographers of diesel locomotives. Information about ALCO locomotives has been sourced from the reference [2], which we gratefully acknowledge.

Appendix: Locomotive Codes

W:	Broad Gauge
Y:	Meter Gauge
Z:	Narrow Gauge, 2 ft 6 in.
N:	Narrow Gauge, 2 ft
D:	Diesel
M:	Mixed Traffic (Both freight and passenger)

G:	Goods Locomotive
P:	Passenger
S:	Shunter
WDM2:	Broad gauge diesel mixed traffic locomotive, model number 2
WDM3A:	Broad gauge diesel of mixed traffic design. 3A represents 3100 hp similarly in WDM3D, 3D represents 3400 hp so it's a slightly strange nomenclature. Here 3 means the 3000 hp class locomotives with A, B, C, D meaning 3100 hp, 3200 hp and so on. WDG4 actually means the goods locomotive and model number 4 by linking it with WDM4, the first broad gauge model from GM but one can also interpret it as 4000 hp
Co-Co:	Two wheel-sets each of three axles, hence six axles in total. Each of the axles has one traction motor
Bo-Bo:	Two wheel set with two axles each and each axle has one traction motor
DC/DC:	DC Generator feeding DC traction motors
AC/DC:	AC Alternator feeding DC traction motors
First BG Diesel shed:	Gaya, homing WDM1
First WDM2 Shed:	Katni, developed with help from ALCO
First EMD Shed:	Hubli

References

1. Martin Terry (2000) Halfway to heaven. Rail Romances Speciality Publishers, UK
2. Sharma SM, Dutta J (2009) The story of Indian ALCOs: legend of the WDM2. Published by the National Rail Museum, New Delhi, India
3. Gautam A, Chandra P, Kumar K, Sharma MR, Kumar S, Agarwal AK (2012) Development of an electronic fuel injection system for a 4-stroke locomotive diesel engine. Proceedings of the ASME 2012 Internal Combustion Engine Division, Spring Technical Conference, ICES 2012, May 6–9, Torino, Pieomonte, Italy

Part II
Efficiency Improvement and Noise Reduction

Exhaust Heat Recovery Options for Diesel Locomotives

Gaurav Tripathi and Atul Dhar

Abstract Even by conservative estimates more than 20% fuel energy from internal combustion engines is wasted as exhaust heat. Currently organic Rankine cycles and thermoelectric generators are most widely investigated options for automobile exhaust heat recovery. Use of thermoelectric generators for recovery of exhaust heat in automobiles at concept level started few decades ago. Major advantages of this technology over Rankine cycles are little noise and vibration, high durability, environmental friendliness, and low maintenance cost for converting low quality thermal energy directly into high quality electrical energy. Major challenges are lower efficiency ($\sim 8\%$), drop in efficiency at lower temperatures, performance optimization in synchronization with multiple constraints of after-treatment devices, silencer, back pressure reduction, turbo-charging etc. Larger size of diesel locomotives compared with space available for automobile engine's mounting on vehicles makes the installation of exhaust heat recovery system in diesel locomotives more practical. In this paper, feasibility and suitability of various exhaust heat energy recovery methods for diesel locomotives has been discussed.

Keywords Exhaust heat recovery · ORC · Thermoelectric · Locomotive

1 Introduction

The exhaust heat recovery (EHR) in diesel engines is very important because of limited energy sources for power production. Extraction of fuel and energy production from fuel are significantly resource intensive activity and EHR saves these resources. Approximately 25–30% of fuel energy is wasted through exhaust gas in various types of small and heavy size engines as shown in Fig. 1. Since locomotive engines power production capacity is large produces hence it has significant heat

G. Tripathi · A. Dhar (✉)
School of Engineering, IIT Mandi, Mandi, India
e-mail: add@iitmandi.ac.in

A.K. Agarwal et al. (eds.), *Locomotives and Rail Road Transportation*,
DOI 10.1007/978-981-10-3788-7_3

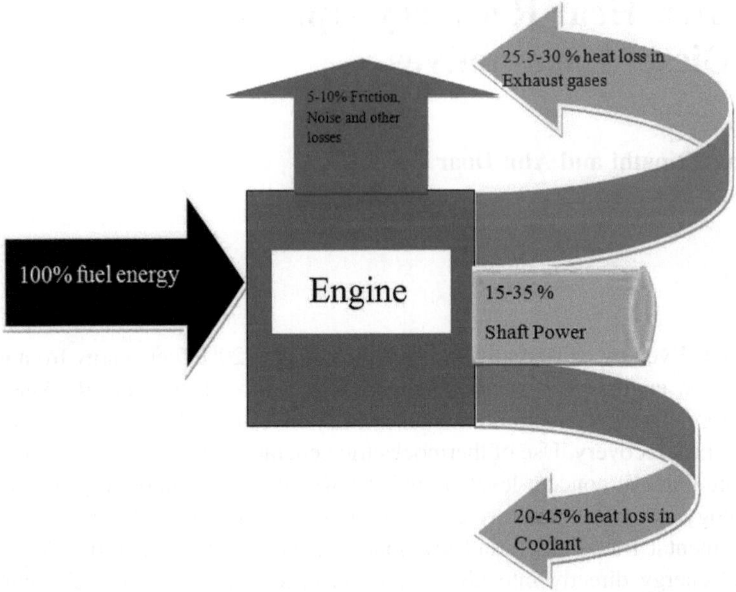

Fig. 1 Sankey-diagram for distribution of fuel energy in typical diesel engine [3, 12]

energy share in exhaust gas therefore viable usable amount of energy can be recovered. Generally in locomotive engine, heat energy is wasted in two different forms. It is wasted from coolant and high temperature exhaust gases. As the temperature of exhaust gas is much higher than average temperature of coolant hence focus is put on EHR from high temperature exhaust gases. Requirement of system installation space is a critical problem associated with EHR system for diesel engine vehicles. Locomotive engines has plenty of space compared to bike, car, truck and other diesel engine vehicles hence they are more suitable for implementation of EHR systems. Another big problem related to EHR system is fluctuation of load due to variation in exhaust gas temperature and quantity. In case of locomotives even this concern is comparatively less severe because it has comparatively defined travelling path as well as limited number of stoppages (correspondingly limited number of operating notches for diesel locomotives) which increases the certainty of exhaust energy content.

List of various EHR methods for recovering exhaust heat from diesel engines is shown in Table 1. To recover exhaust heat from locomotive engine (a) Fluid base EHR approach and (b) Solid state base EHR are there are mainly two available approaches. This paper deals with parameters related with EHR system such as efficiency of different methods, space requirement, vibration and noise problems, design modification problem etc. And try to conclusive comparison among various available EHR systems for diesel locomotive railway engine.

Table 1 Various methods of engine EHR

S. No.	EHR method	Engine type	Efficiency or performance	References
1	Thermal system with Rankine cycle	Road vehicle or off road vehicle	Efficiency up to 30%	[1]
2	Rankine cycle with axial swash plate expander	Passenger car	Cycle efficiency up to 13%	[2]
3	Waste heat recovery method based on Rankine cycle (RC) Desalination Combined cycle Refrigeration system	 Ship Ship Ship Ship	By the use of RC overall efficiency increased by 8.03% and in some cases 11.4%	[3]
4	Organic Rankine cycle with a single screw expander	Diesel exhaust	ORC efficiency is 6.48%	[4]
5	Large fuel cell	Locomotive	Increase in overall efficiency approximately 19%	[5]
6	Absorption ice maker	Boat	Optimum COP is 0.125	[6]
7	Cabin air conditioning system	Railways locomotive	COP is too low 0.2 (but rated COP is 0.2–0.3)	[7]
8	Cabin air conditioning system	Car, bus, truck and train	Optimum COP is 0.39	[8]
9	Waste heat driven jet ejector cooling system of gasoline engine	Gas turbine	Thermal COP is 0.07–0.26 and hydraulic COP is 6.4–29.3	[9]
10	Vapour absorption refrigeration system	Automobile	At 1000 rpm, minimum generator temperature is 75 °C and maximum evaporator temperature is 20 °C	[10]
11	Waste energy and solar energy combination is used by absorption system for chilling water, air conditioning and ice making	Diesel engine exhaust and power plant exhaust	–	[11]

2 Fuel Based Exhaust Recovery Approach

Fluid based EHR approach uses a Rankine cycle for recovering heat energy from engine exhaust. In Rankine cycle, fluid is circulated in cycle to extract the heat from coolant or from exhaust of locomotive. Thermal system with Rankine cycle is most efficient and less costly system. Use of EHR method depends on size and power of engine and also depend on the fact that vehicle is stationary powertrain, road vehicle or off road vehicle. EHR method has problem in dealing with on road vehicle because it does not produce high temperature generally and it has space problem for large EHR hardware [1]. Two types of Rankine cycles are used for fluid base EHR systems namely (a) Steam Rankine cycle (SRC) and (b) Organic Rankine cycle (ORC). In steam Rankine cycle, working fluid of Rankine cycle is water, which exchanges heat with engine exhaust and gets evaporated and then expands on turbine to produce power. But there is problem of dissolved O_2 and CO_2 gases in steam that needs equipment to remove dissolved gases. Yamada et al. reported that use Rankine cycle with hydrogen internal combustion engine resulted in increase in thermal efficiency. They used steam as a working fluid for Rankine cycle. In open steam based Rankine cycle without condenser they observed 2.9–3.7% increase in overall thermal efficiency in the engine operating speed range of 1500–4500 rpm [13].

In organic Rankine cycle instead of water, organic fluid is used as working fluid, which has many benefits in comparison of steam Rankine cycle such as low operating pressure hence lesser vibration problem, capability to operate with low temperature source, efficiency compared to SRC etc. In terms of use of exhaust heat, ORC is better than steam Rankine cycle. Chammas et al. reported that with ORC there is more improvement in fuel economy and higher power output in comparison to SRC [14]. Srinivasan et al. reported that use of organic Rankine cycle with advance injection low pilot ignition Natural gas engine shows that there is improvement in fuel conversion economy is 10% at half load and reduction in NO_x [15].

There are various options in implementation of ORC based EHR systems. In ORC that directly run on coolant without boiler heating, heated coolant outlet from engine is send to ORC and directly expanded on turbine and produces electricity. Low temperature coolant is send back to engine. Another similar implementation of ORC that directly works on coolant with boiler heating. In these systems, coolant from engine exhaust is send to ORC boiler. In boiler, heated coolant is again heated and expanded on turbine and produces electricity. After this low temperature coolant is send back to engine. In some ORC systems coolant or exhaust gases are used to preheat the organic fluid. In these systems, coolant is used to preheat the organic fluid. Then preheated organic fluid is again heated in evaporator. After this evaporated organic fluid is expanded on turbine and produces power. ORC can be used with preheater and evaporator, in both cases thermal efficiency of ORC increases. Preheater is used to preheat the working fluid with the heat of engine-out coolant. Evaporator is used to heat working fluid with the help of engine exhaust. Recuperator can also be used in place preheater, which preheats the working fluid with the help of released heat from condenser. More increase in efficiency of RC is observed with use of recuperator in place of preheater [16].

Power recovered in ORC systems can be used either for increasing power output of engine or generating the electricity. In first system, exhaust of engine is utilized by ORC to evaporate the working fluid and evaporated fluid is allowed to expand on turbine and produces power. Then ORC turbine output power is coupled with engine output power and in this way net power output from engine increases (Fig. 2).

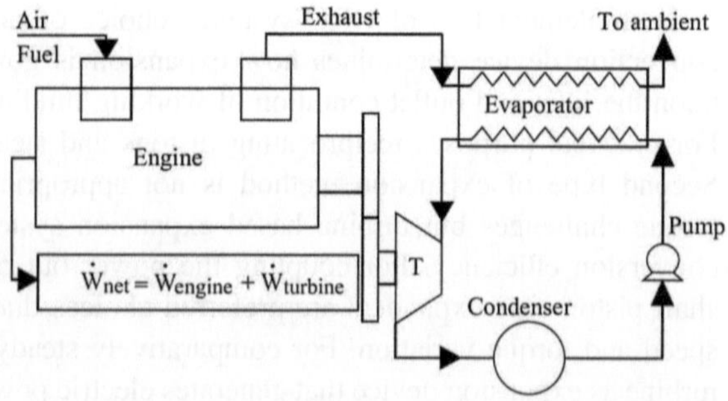

Fig. 2 ORC for increasing power output of engine [17]

Figure 3a, b explain the working of ORC for generating electricity. In these systems, exhaust energy is utilized by ORC to evaporate the working fluid and evaporated fluid is allowed to expand on turbine and produces power. That power is used for generating electricity with the help of generator. In Fig. 3a, preheating of working fluid is done with engine coolant and evaporation of working fluid is done with engine exhaust. While in Fig. 3b, preheating of working fluid is done with recuperator with the help of heat rejected during expansion process. Heat for the evaporation of working fluid is supplied from engine exhaust. Both arrangements result in comparable increase in system efficiency [18].

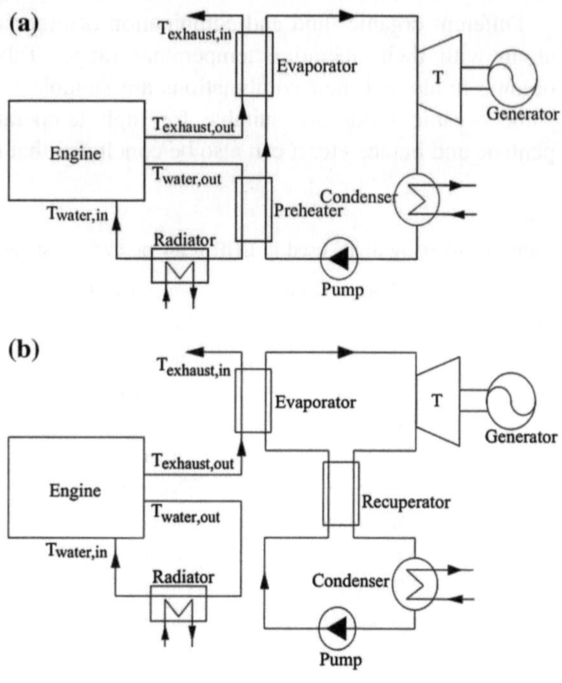

Fig. 3 ORC for generating electricity from engine exhaust: **a** with preheater and **b** with recuperator [18]

In implementation of ORC systems, choice of pressure to mechanical motion conversion device determines how expansion is going to take place. It depends upon the inlet and outlet condition of working fluid and size and output of engine. For practical purposes reciprocating pistons and steam turbines are used options. Second type of expansion method is not appropriate to deal with velocity and torque challenges but turbine based expansion system has better thermal energy conversion efficiency. For coupling the power out of Rankine cycle with engine shaft piston type expanders are preferred choices due to their capability to handle speed and torque variation. For comparatively steady exhaust energy input use of turbine as expansion device that generates electric power is consider as better option [12].

Two different types of working fluids are used in RC that are inorganic fluid and organic fluid as shown in Table 2. Inorganic fluids are also known as wet fluid because expansion process occurs in wet region. Ammonia and water are example of inorganic fluids. Generally, inorganic fluid is used in SRC. There are some benefits in operating with inorganic fluid such as RC can operate at high temperature and high thermal efficiency. Disadvantages in dealing with inorganic fluids include erosion of blades, requirement of reheating and superheating etc. So it can be concluded that inorganic fluids and SRC are suitable for power plant application not for automobile EHR system. Organic working fluids are also known as dry or isentropic fluid because expansion process occurs in dry region and very close to isentropic expansion [19]. Wet fluids such as R-11 organic solvent are more suitable for low temperature EHR operations. Organic fluids also offer better part load efficiency and engine coolant can be used as working fluid.

Different organic fluid and combination of organic fluids are shown in Table 3 along with their operating temperature range. Table 3 concludes that most of organic fluids and their combinations are suitable for low temperature range. But some organic fluids are suitable for high temperature range such as benzene, pentane and butane etc. It can also be concluded that mixture of organic fluids such

Table 2 Working fluids used in EHR Rankine cycle systems [20]

S. No.	Working fluid	Category of fluid	Type of working fluid
1	Water	Inorganic	Wet
2	Ammonia	Inorganic	Wet
3	Ethanol	Inorganic	Wet
4	R11	Organic	Isentropic
5	R123	Organic	Isentropic
6	Benzene	Organic	Isentropic
7	n-Pentane	Organic	Dry
8	Iso-Pentane	Organic	Dry
9	Toluene	Organic	Dry
10	p-Xylene	Organic	Dry
11	HFE7 100	Organic	Dry

Table 3 Dependency of ORC fluid on temperature

S. No.	Organic fluid of RC	Temperature	Efficiency	References
1	R123	373–453 K	Optimum	[21]
2	R141b	>453 K	Optimum	[21]
3	Any organic fluid	<373 K	Not used because not economical	[21]
4	R-245fa, R113, R123	313–473 K	R-245fa is efficient than other two fluid	[22]
5	R134a	433–573 K	15–17% increase in efficiency at maximum temperature	[23]
6	HCFCs-141b	368.4–369.1 K	Thermal efficiency is 0.013–0.052%	[24]
7	Quaternary mixture refrigerant (mixture of R-125/R-123/R-124/R-134a)	<= 338 K or 338–755 K	Quaternary mixture has better exergy efficiency and 2nd law efficiency than R-245fa, R-11, R-114	[25]
8	R134a, R141b	373, 438 and 503 K	R141b has higher efficiency and lower cost than R134a at 230 °C	[26]
9	Mixture of R245fa and R152a	343–373 K	80% of R245fa and 20% R152a mixture has high first law and second law efficiency at all temperature	[27]
10	Benzene, n-pentane, Iso-pentane, R123, Butane, R245fa	373–673 K	Benzene > n-pentane > Iso-pentane > R123 > Butane > R245fa shows variation of thermal efficiency	[28]
11	R11, R123, R113, R114, R245fa, R236fa	380.37–644.26 K	R11 > R123 > R113 > R114 > R245fa > R236fa variation of thermal efficiency	[29]
12	R11, R123, R113, R114, R245fa, R236fa, R134a	338.7–380.37 K	R11 < R123 < R113 < R114 < R245fa < R236fa < R134a variation of thermal efficiency	[29]
13	R123, R245ca, R245fa, Butane, R236ea, R142b, Isobutene, R236fa, R124, R152a, R227ea, R134a, Propylene, R32, R143a, R218, R125, R41	340–540 K	R152a and R143a have maximum output and minimum investment	[30]
14	R134a and isobutene	393 K	R134a and Isobutene have exergy efficiency 19.6 and 20.3%	[31]
15	Toluene, cyclohexane, R123, R245fa	393–523 K	Overall efficiency Toluene > Cyclohexane > R123 > R245 fa	[32]

as such as mixture of R245fa and R152a and quaternary mixture refrigerant have high efficiency in comparison to single operating organic fluid.

3 Solid-State Based EHR

Solid state based EHR uses thermo-electric generator (TEG), which converts the exhaust energy passing through it into electricity through Seebeck effect. Basic building blocks of TEGs are referred a thermo-electric modules (TEMs). The thermo-mechanical stress is a problem related with TEG which decreases the TEG performance. Thermo-mechanical stress occurs because of temperature gradient in the TEG. Another problem related with thermo-mechanical stress is that it has contradictory nature with electric output. Investigation by Heghmanns et al. shows that with double sided bondable and flexible thermoelectric module, thermo-mechanical stress reduces and electric power increases without exceeding the limit of maximum exhaust back pressure [33].

Availability of TEM materials, which have higher thermal to electrical energy conversion efficiencies in the temperature range of exhaust gases are critical for the efficiency of these systems. Table 4 shows comparison among various TEG materials with the help of ZT values. ZT value is known as figure of merit value. ZT value is a dimensionless number that is used to determine ability to produce electricity from available heat source. ZT value is given by;

$$ZT = \frac{\sigma S^2 T}{K}$$

where σ is electric conductivity, S is Seebeck coefficient and K is thermal conductivity and T is temperature. ZT value and efficiency of these materials is highly dependent on temperature and it drops drastically with deviation of temperature from the optimum value. Variation in exhaust gas temperature with locomotive operating conditions is the major cause of overall low efficiency of TEG systems.

Apart from choice of suitable material, factors such as compressive load on TEM, thermal expansion of module arrays, uniform optimum loading on different modules, uniform heating of TEM surfaces, interception of generated heat by TEM affects the efficiency of TEG based EHR systems at component level implementation. In presence of compressive load on TEG, performance of TEG enhances as there is increase in surface contact of TEG with hot and cold surface. Hence, it is preferred to exert some compressive load in the installed position of TEMs. TEG module operates at high temperature so there are chances of thermal expansion of bolt and threads, which result in decrease in performance with time. So spring is used to maintain compressive load on TEG which also eliminates the problem of thermal expansion [41].

Application of large clamping force on TEG results in non-uniform loading. In extreme case, application of large clamping force result in crack formation in TEG.

Table 4 Comparison among various TEG performances

S. No.	Material of TEG	Temperature	Remark	References
1	Alloys of Bismuth (Bi) with Anitmony (An) and Tellurine (Te) or Selenium (Se)	450 K	ZT is up to 2.5, expensive	[34]
2	Alloys of Lead (Pb)	850 K	ZT is up to 2.5 expensive	[34]
3	SiGe alloys	1300 K	ZT is up to 2.5 expensive	[34]
4	n-type MgSi based alloy	723 K	ZT is approximately 1.1	[35]
5	Synthesizing $Mg_2SiMg_2SnMg_2Ge$	800 K	ZT is approximately 1.08	[35]
6	Mg_2Si nano-composites	773 K	ZT is approximately 0.7	[35]
7	Bi_2-Te_3	523 K	ZT is approximately 1.07	[36]
8	p-type-PbTe nanostructured with SrTe	915 K	ZT is approximately 2.2	[37]
9	Skutterudite (squares, data depicted for $Ba_{0.08}La_{0.05}Yb_{0.04}Co_4Sb_{12}$)	915 K	ZT is approximately 2.2	[38]
10	Tetrahedrite (diamonds, data depicted is for 0.5 natural mineral $Cu_{9.7}Zn_{1.9}Fe_{0.4}As_4S_{13}$ and 0.5 synthetic $Cu_{12}Sb_4S_{13}$)	723 K	ZT is approximately 1	[39]
11	Half-Heusler (triangles, data depicted for $Nb_{0.6}Ti_{0.4}FeSb_{0.95}Sn_{0.05}$)	–	–	[40]

To remove this problem thickness of hot and cold surfaces is increased. If there is temperature difference across the TEG surface due to different temperate at edges then the performance of TEG decreases. To remove this problem, some part of TEG is over hanged [41]. To achieve uniform temperature across hot and cold junctions, a thick (0.25 in.) copper or aluminium sheet is inserted between TEG and hot source and another copper or aluminium sheet between TEG and cold source.

To increase the heat transfer between TEG and hot body or cold body, proper contact is required. So surfaces are made flat and voids between surfaces are filled with some heat transfer compound. The performance of TEG decreases if heat is transferred from hot body to cold body without passing through TEG. This is known as thermal bypass. To decrease thermal bypass energy, it is necessary to make clamping materials as small as possible and decrease their thermal conductivity [41].

Table 5 Efficiency comparison of various EHR methods [42]

S. No.	EHR method	Efficiency (%)	References
1	TEG with locomotive radiator	2	[42]
2	TEG with locomotive exhaust	<5	[42]
3	Steam Rankine cycle	10–20	[42]
4	Organic Rankine cycle	17–20	[42]
5	ORC operated with coolant preheated working fluid and boiler	8–10	[42]
6	ORC operated with coolant as working fluid without use of boiler	4	[42]
7	ORC operated with boiler and coolant as working fluid	6	[42]
8	ORC output coupled with engine output	10	[17]

4 Comparison of Solid and Fluid Based EHR Option

For typical EHR conditions, more power output and fuel economy is achieved in case of ORC in comparison to high and low pressure SRC. Efficiency comparison of ORC at various engine operating points indicates that thermal efficiency of ORC is higher at lower engine loads in comparison to higher loads. Efficiency of ORC is increased by use of preheater/recuperator that preheats the working fluid with the help of engine coolant or with heat release from condenser. Construction cost, ease of operation and maintenance favours TEG but their energy conversion efficiency is comparatively lower than ORC. Table 5 shows the efficiency comparison among various solid and fluid based EHR methods and indicates that ORC option is best in comparison to other available EHR options.

Table 6 Thermal properties of good ORC fluid [43]

S. No.	Properties	Desired range
1	Boiling point	323–473 K
2	Melting point	<233 K
3	Ignition point	>473 K
4	Heat capacity	Low
5	Thermal conductivity	High
6	Enthalpy of evaporation	High
7	Thermal or chemical breakdown	High as possible
8	Decomposition temperature	High as possible

5 Case Studies on EHR Method Implementation

Automobile research company FEV North America and FEV Germany has done a lot of research on ORC fluid and provide specific properties that are given in Table 6. Along with these properties ORC fluid should be compatible with sealing material, steel, aluminium and ORC system. And ORC fluid should have optimum level of critical pressure, temperature and vapour pressure. And FEV provide some ORC fluid such as ethanol, acetone, toluene and R245fa which has less production cost, easy availability and large temperature and pressure range.

In ThermaDynamics Rail LLC EHR method implemented on locomotive [44], a thermodynamic cycle is used for extracting exhaust gas energy of engine and converted into mechanical or electric output as shown in Fig. 4. In this process, working fluid is pressurised with the help of pump and supplied into High Pressure Heat Exchangers (HiPHEXs). HiPHEXs exchanges heat between engine exhaust and working fluid. Superheated working fluid expands on turbine and produces mechanical shaft power. That can be further converted into electric power with the help of Power Conversion Unit (PCU). For completing cycle, a condenser is used that converts vapour into liquid state. This condensed working fluid is available at pump inlet. Pump supplies this working fluid at high pressure to HiPHEXs to complete the cycle. Electric output varies according to locomotive operating conditions. It is found that 11.8% power can be recovered.

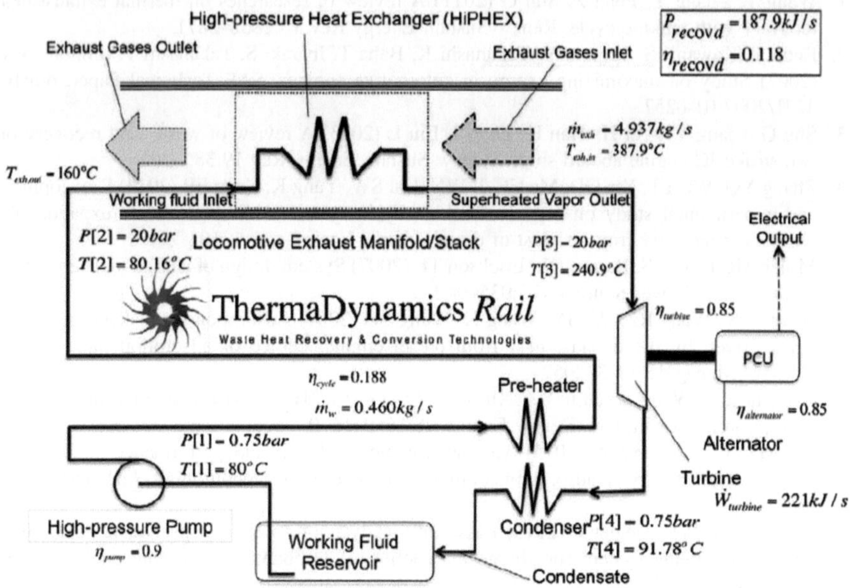

Fig. 4 Components of TDR EHR method [44]

6 Summary and Possibilities

For stationary power plants and constant load application EHR system is already available in commercial form but there is problem in application of EHR system in locomotive and other diesel engines. This problem can be removed by efficient design modification of EHR system as well as by carrying out necessary design modifications in locomotives and other diesel engines. This can be done by more encouraging research and development of locomotives and EHR system.

Although the efficiency of ORC is highest among the available EHR methods, but there is need to find suitable working fluid of ORC for large temperature ranges. TEG is compact EHR option but there is need to increase the conversion efficiency by searching suitable material and enhance heat transfer coefficient of heat exchanger on both cold and hot side.

Regarding further development possibilities, it can also be suggested that instead of using one EHR system, use primary and secondary EHR system. Primary EHR system based on Steam Rankine cycle or Organic Rankine cycle for recovering heat energy from engine exhaust and secondary EHR system for recovering energy of primary EHR system. Combination of primary and secondary EHR system may provide better efficiency.

References

1. Wang T, Zhang Y, Peng Z, Shu G (2011) A review of researches on thermal exhaust heat recovery with rankine cycle. Renew Sustain Energy Rev 15:2862–2871
2. Endo T, Kawajiri S, Kojima Y, Takahashi K, Baba T, Ibaraki S, Takahashi T, Shinohara M (2007) Study on maximizing exergy in automotive engines. SAE Technical Paper. doi:10.4271/2007-01-0257
3. Shu G, Liang Y, Wei H, Tian H, Zhao J, Liu L (2013) A review of waste heat recovery on two-stroke IC engine aboard ships. Renew Sustain Energy Rev 19:385–401
4. Zhang YQ, Wu YT, Xia GD, Ma CF, Ji WN, Liu SW, Yang K, Yang FB (2014) Development and experimental study on organic Rankine cycle system with single-screw expander for waste heat recovery from exhaust of diesel engine. Energy xxx 1–10
5. Miller AR, Hess KS, Barnes DL, Erickson TL (2007) System design of a large fuel cell hybrid locomotive. J Power Sources 173:935–942
6. Wang LW, Wang RZ, Wu JY, Wang K, Wang SG (2004) Adsorption ice makers for fishing boats driven by the exhaust heat from diesel engine: choice of adsorption pair. Energy Convers Manag 45:2043–2057
7. Jiangzhou S, Wang RZ, Lu YZ, Xu YX, Wu JY, Li ZH (2003) Locomotive driver cabin adsorption air conditioner. Renew Energy 28:1659–1670
8. Ali MS, Chakraborty A (2015) Thermodynamic modeling and performance study of an engine waste heat driven adsorption cooling for automotive air-conditioning. Appl Therm Eng 90:54–63
9. Zegenhagen MT, Ziegler F (2015) Feasibility analysis of an exhaust gas waste heat driven jet-ejector cooling system for charge air cooling of turbocharged gasoline engines. Appl Energy 160:221–230

10. Rego AT, Hanriot SM, Oliveria AF, Brito, Rego TFU (2014) Automotive exhaust gas flow control for an ammonia-water absorption refrigeration system. Appl Thermal Eng 64:101–107
11. Wang RZ, Oliveira RG (2006) Adsorption refrigeration—an efficient way to make good use of waste heat and solar energy. Prog Energy Combust Sci 32:424–458
12. Stobart R, Weerasinghe R (2006) Heat recovery and bottoming cycles for SI and CI engines—a perspective. In: SAE paper 2006-01-0662
13. Yamada N, Mohamad MNA (2010) Efficiency of hydrogen internal combustion engine combined with open steam Rankine cycle recovering water and waste heat. Int J Hydrogen Energy 35:1430–1442
14. Chammas RE, Clodic D (2005) Combined cycle for hybrid vehicles. In: SAE paper 2005-01-1171
15. Srinivasan KK, Mago PJ, Zdaniuk GJ, Chamra LM, Midkiff KC (2008) Improving the efficiency of the advanced injection low pilot ignited natural gas engine using organic Rankine cycles. J Energy Resour Technol Trans ASME 130:0222011–7
16. Vaja I, Gambarotta A (2010) Internal combustion engine (ICE) bottoming with organic Rankine cycles (ORCs). Energy 35:1084–1093
17. Srinivasan KK, Mago PJ, Zdaniuk GJ, Chamra LM, Midkiff KC (2008) Improving the efficiency of the advanced injection low pilot ignited natural gas engine using organic Rankine cycles. J Energy Resour Technol Trans ASME 130:0222011an
18. Vaja I, Gambarotta A (2010) Internal combustion engine (ICE) bottoming with organic Rankine cycles (ORCs). Energy 35:1084
19. Chen SK, Lin R (1983) A review of engine advanced cycle and Rankine bottoming cycle and their loss evaluations. In: SAE paper 830124
20. Liu BT, Chien KH, Wang CC (2004) Effect of working fluids on organic Rankine cycle for waste heat recovery. Energy 29:1207–1217
21. Wang ZQ, Zhou NJ, Wang XY (2012) Fluid selection and parametric optimization of organic Rankine cycle using low temperature waste heat. Energy 40:107–115
22. Li Y (2012) Analysis of low temperature of organic Rankine cycle for solar applications. Lehigh University
23. Ko HJ, Kim SW, Han CH, Kim KH (2013) Effects of source temperature on thermodynamic performance of transcritical organic cycle. Int J Mater Mech Manuf 1(1)
24. Saiai P, Chaitep S, Bundhurat D, Watanawanyoo P (2014) Effect of vapor generator on organic Rankine cycle for low temperature heat source. IJETAE 4(1)
25. Sami SM (2008) Energy and exergy analysis of an efficient organic Rankine cycle for low temperature power generation. Int J Ambient Energy 29(1)
26. Khennich M, Galanis N (2012) Optimal design of ORC system with a low temperature heat source. Entropy 14:370–389. doi:10.3390/e14020370
27. Deethayat T, Kiatsiriroat T (2015) Performance analysis of an organic Rankine cycle with internal heat exchanger having zeotropic working fluid. Case Stud Thermal Eng 6:155–161
28. Adhouri M, Ahmadi MH, Feidt M (2014) Performance analysis of organic Rankine cycle integrated with a parabolic through solar collector. In: World Sustainability Forum 2014—Conference Proceedings Paper
29. Brasz LJ, Bilbow WM (2004) Ranking of working fluids for organic Rankine cycle applications. In: International refrigeration and air conditioning conference, Purdue University
30. Gao H, Liu C, He C, Xu X, Wu S, Li Y (2012) Performance supercritical organic Rankine cycle for low grade waste heat recovery. Energies 5:3233–3247. doi:10.3390/en5093233
31. Darvish K, Ehyaei MA, Atabi F, Rosen MA (2015) Selection of optimum working fluid for organic Rankine cycle by exergy and exergy-economics analyses. Sustainability 7: 15362–15383. doi:10.3390/su71115362
32. Wang X, Yang Y, Wang M, ZhengYa, Dai Y (2015) Utilization of waste heat from intercooled reheat and recuperated gas turbines for power generation in organic Rankine cycles. Research Gate, Paper ID 28, p 1

33. Heghmanns A, Beitelschmidt M, Wilbrecht S, Geradts K, Span G (2015) Development and optimization of a TEG-system for the waste heat usage in railway vehicles. Mater Today Proc 2:780–789
34. Patil D, Arakerimath RR (2013) A review of thermoelectric generator for waste heat recovery from engine exhaust. IJRAME 1(8):1–9
35. Fairbanks J (2013) Automotive thermoelectric generator and HVAC. Sustainable Transportation, US department of Energy, Energy Efficiency and Renewable Energy
36. Ramade P, Patil P, Shelar M, Chaudhary S, Yadav S, Trimbake S (2014) Automobile exhaust thermo-electric generator design and performance analysis. IJEATE 4(5)
37. Biswas K, He J, Blum ID, Wu CI, Hogan TP, Seidman DN, Dravid VP, Kanatzidis MG High performance bulk thermoelectrics with all-scale hierarchical architectures. Letter. doi:10.1038/nature11439
38. Shi X, Yang J, Salvador JR, Chi M, Cho JY, Wang H, et al (2011) Multiple-filled skutterudites: high thermoelectric figure of merit through separately optimizing electrical and thermal transports. J Am Chem Soc 133(20):7837–7846
39. Lu X, Morelli DT (2013) Natural mineral tetrahedrite as a direct source of thermoelectric materials. PhysChemChemPhys 15(16):5762–5766
40. Joshi G, He R, Engber M, Samsonidze G, Pantha T, Dahal H et al (2014) NbFeSb-based p-type half-Heuslers for power generation applications. Energy Environ Sci 7:4070–4076
41. Leavitt FA, Elsner NB, John C Use, application and testing of Hi-Z thermoelectric modules (The Hz-14 is used as an example. The other modules should be evaluated in a similar way.) Bass Hi-Z Technology, Inc
42. Francesco S, Juergen P (2010) Enhanced locomotive efficiency through waste heat recovery. In: Conference on railway engineering wellington, 2010
43. Jeihouni Y, Franke M, Lierz K, Tomazic D, Heuser P (2015) Waste heat recovery for locomotive engines using the organic Rankine cycle. In: Proceedings of the ASME 2015 internal combustion engine. In: Division Fall Technical Conference ICEF2015, November 8–11, 2015, Houston, TX, USA
44. Filippone C (2014) Diesel-electric locomotive energy recovery and conversion. innovations deserving exploratory analysis (IDEA) programs managed by the Transportation Research Board (2014)

Diesel Locomotive Noise Sources, Reduction Strategies, Methods and Standards

Nachiketa Tiwari

Abstract Diesel locomotives are the mainstay of modern railway systems. Despite the fact that they are noisy and polluting compared to electric locomotives, they continue to remain popular across the world. This is because they can run in areas where electric supply is either unavailable or unreliable, and also because in times of war their service does not get affected even if electric power lines get damaged. Since most railway networks cut across highly populated areas, efforts have to be made to ensure that diesel engine noise is reduced to a minimum. In this chapter this need has been addressed. Specifically, the chapter explains the working of a modern diesel engine locomotive in brief. Such an understanding is important, as all noise generating mechanisms in a diesel locomotive are tightly coupled with its working. Next, different types of noise emanating from a diesel locomotive have been described. Such a description will help the reader assess the nature of each noise component in terms of its magnitude and its frequency spectrum. Next a method to develop noise reduction strategy of diesel locomotive noise is proposed. Next, the chapter explains a diverse suite of noise reduction methods. These methods can be passive or active. The range of passive methods is fairly large. All these methods have been described in detail. This is followed up with a description of important regulations related to locomotive noise. Finally, an exhaustive list of relevant international standards has been provided. These references will be very useful to the reader if she is involved with reduction of train and locomotive noise.

1 Background

Diesel locomotives are the mainstay of railway networks worldwide. Despite the fact that they pollute air, produce significant levels of noise, and consume excessive amounts of oil, they have not been replaced by electric locomotives because of

N. Tiwari (✉)
Department of Mechanical Engineering, Indian Institute of Technology Kanpur, Kanpur, India
e-mail: ntiwari@iitk.ac.in

A.K. Agarwal et al. (eds.), *Locomotives and Rail Road Transportation*,
DOI 10.1007/978-981-10-3788-7_4

three reasons. One, they can run in areas where electricity supply is either unavailable, or unreliable. Two, they are used for strategic reasons so that critical supply lines do not get affected even if electric power lines get damaged in times of war or other exigencies. Three, they are one of the most efficient means to convert chemical energy locked in diesel into electricity. Thus, while there are about 37,000 electric locomotives in operation today [1], the number of diesel locomotives currently in operation is much more and is estimated to be 120,000 [2]. This number is growing each year. The global market for new diesel locomotives is estimated to be around 4.8 billion Euros [3]. With an average diesel locomotive costing approximately 1.7 million dollars [4], the number of in-operation diesel locomotives across the world is growing by 3,300 each year. This annual increment is expected to increase each year with the growth of BRICS (Brazil, Russia, India, China and South Africa) and other emerging economy countries. Thus, it is becoming increasingly important to understand noise generating mechanisms of diesel locomotives, and also develop methods to reduce the same.

The first compression ignition (CI) engine was developed by Rudolf Diesel in 1892. Prior to that, gasoline based IC engines were used to power locomotives. Initial attempts at using IC engines (CI and SI) were not all that successful. The underlying reason was that IC engines perform optimally only within a narrow RPM band. This limitation was addressed through use of clutches and gearboxes. However, such an approach works only in light applications, such as cars and trucks. For super-heavy applications clutches do not work, as they wear down too soon. Further, a typical IC engine for automotive applications runs between 500 and 6000 RPM. In contrast the working RPM for a diesel locomotive engine is approximately between 269 and 904 RPM. With such a very narrow range of RPMs, the engine system would need a very large gear box, perhaps having dozens of gear-sets so that the train could run at high speeds. Such a solution is impractical because of cost, mass and size considerations. Due to these limitations, there was little progress in using diesel engines for locomotive applications in the initial years.

Later this limitation was overcome through development of the diesel-electric combination drive in the decade of 1920s. In such systems the diesel engine would drive an electrical generator, and the power from the generator would run electrical motors to generate traction. Sophisticated electronic control systems ensured that the efficiency of traction motors did not degrade due to changes in its running RPM. These developments eliminated clutch, and gear box based drive systems for a diesel locomotive. The diesel-electro drive concept capitalized on this feature of electric motors, and paved the way for large scale use of diesel engines for locomotive applications. It is not that all diesel locomotives are based on the diesel-electric drive concept. There are other concepts as well. However, the diesel-electric concept is much more popular, and is almost universally used nowadays. Such diesel locomotives offer much greater flexibility and performance than the steam engine. Diesel locomotives, unlike steam locos, can be almost instantly turned on and off. Compared to a steam engine, a diesel engine is quieter, weatherproof, and offers a cleaner and better work environment for the engine operator. Further, several of these engines can be used in a single train.

However, this option is not practical in case of steam engines. This feature brings greater efficiencies to the railway operator. It also makes possible the use of several standard low-power diesel locomotives for hauling long and heavy trains. Steam engines require large quantities of coal and water, and their efficiency is considerably less that of diesel engines. These engines also have significantly higher operating and maintenance costs. For these reasons, diesel engines have more or less totally replaced steam engines across the world.

However, from the standpoint of pollution, diesel engines, very much like steam engines, offer significant challenges. They emit large amounts of emissions and are very noisy. Since most railway networks cut across areas with high population densities, diesel locomotives expose a large number of human lives to high sound levels. In a study [5] the level of pass-by noise due to a moving diesel train, measured at a distance of 16.5 m from the track, was found to be as high as 115 dB (A). Given the omnipresence of diesel locos, and the proximity of their operational areas to regions of human habitat, it is very important to invest time and resources for reducing diesel locomotive noise. A credible noise reduction strategy is based on three key elements; understanding of noise generation mechanisms in the source, ability to select appropriate noise reduction methods, and putting in place appropriate noise monitoring and control methods.

2 Working of a Diesel Engine

To develop a robust noise reduction strategy for diesel locomotives, it is important to understand the working and construction of a diesel engine. This will help in identifying different noise sources, and their corresponding noise generating mechanisms. There are several types of diesel engine drives. These include diesel-mechanical, diesel-electric, electro-diesel, diesel-hydraulic, diesel-steam, and diesel-pneumatic. Amongst these, the diesel-electric drive is the most common. For this reason, diesel engines are often termed as diesel–electric locomotives. These locomotives work on a simple concept. The engine burns diesel to produce mechanical power. While doing this, diesel engine operates at optimum RPM to maximize on engine efficiency. The mechanical power generated is used to run an alternator or a generator for producing electricity. This electricity is used to run traction motors, which are connected to the axle of locomotive engine. Such a process to convert chemical energy locked in diesel fuel to locomotive motion is more efficient vis-a-vis the process used in steam engines.

Diesel-electric and electric locomotives are similar in two ways. One, they have an electronically controlled electric drive in form of traction motors which drive axles. Two, both types of locomotive systems have an auxiliary system to support on-board needs of electricity required for diverse purposes including lighting, heating, cooking, cooling, and braking. Figure 1 is a schematic representation of a diesel-electric locomotive which is described further.

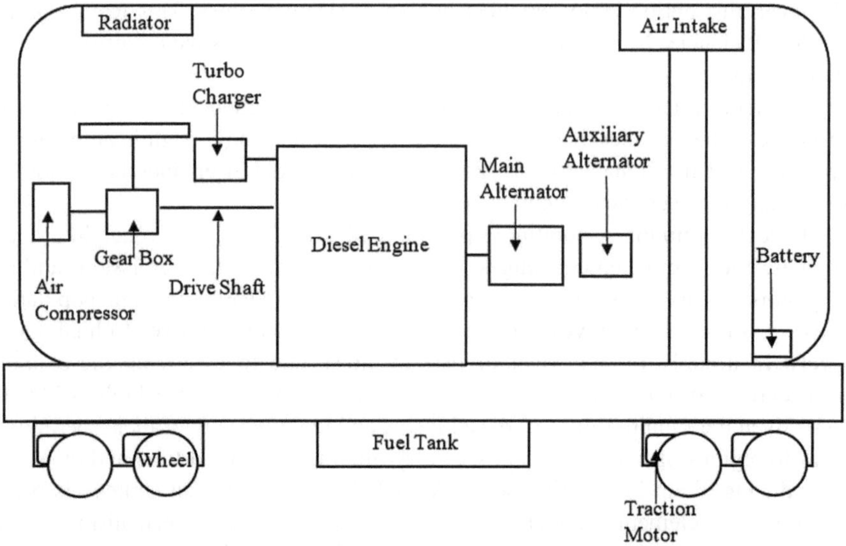

Fig. 1 Schematic representation of a diesel-electric locomotive

2.1 Diesel Engine

Diesel engine is the primary source of energy for the diesel-electric locomotive. It has a large number of cylinders arranged in either a straight line, or a V-configuration, or a "boxer" configuration. These cylinders turn the drive shaft at RPMs rarely exceeding 1200. The drive shaft is connected to either a DC generator, or an alternator. These devices convert mechanical energy into electricity.

2.2 DC Generator or AC Alternator and Rectifier

The mechanical energy generated by a diesel engine is used to generate electricity. For this purpose, DC generators, or electrical alternator-rectifiers are used. In either of these configurations, the system produces DC electricity. Typically, the former option is used for engines providing less than 2200 kW of traction power. For higher power applications, the latter option is used [6]. In either case, there is no mechanical coupling between the diesel engine and locomotive wheels. Earlier, most of the locomotives used DC machines. However, that entailed frequent maintenance as brushes and commutators would wear out rapidly. With the invention of silicon-based rectifiers, these generators have given way to AC generators. The electrical output from these generators is converted into DC electricity through rectifiers. If the engine has a DC generator, then its output is directly converted into 3-phase AC power supply and fed to traction motors. If the system uses AC generators, then the DC output from rectifiers is converted into 3-phase AC power supply to run traction motors.

In addition to the main generator which drives traction motors, diesel engines also have auxiliary generators. These units are used to support on-board electricity needs including those for lighting, heating, cooking, cooling, and braking.

2.3 Traction Motors

The electricity generated from generators is used to traction motors. These motors are connected to axles of the locomotive. The first generation traction motors were of DC type. These motors were preferred over AC motors because their speed-torque characteristics were better than those for contemporary AC motors for driving the train [7]. Specifically, these motors can provide high torque at lower speeds, and thus can provide sufficiently high acceleration to the train. At higher speeds, these motors provide lesser torque. These motors have a stator, which provides the magnetic field, and a rotor which is mechanically attached to locomotive axles. By arranging the field winding with multiple tap points, its speed characteristic can be modulated, thereby facilitating smoother acceleration. However, these motors use brushes and commutators, which require frequent maintenance. For this reason, significant R&D efforts were invested in replacing these motors with AC traction motors which do not use commutators. Their limitation, that they perform optimally only over a narrow speed range, has been addressed through development of Variable Voltage Variable Frequency electronic control systems. The modern diesel-electric locomotive uses four to six 3-phase AC motors. Each of these motors can provide as much as 1,000 hp of power to wheels.

2.4 Electronic Controls and Batteries

All modern diesel locomotives are equipped with an electronic control system. This system is usually located in a cubicle. Unlike automotive engines, the throttle position in locomotive engines is not continuously varying. Rather, it can take a limited number of discrete positions, each position referred as a "notch setting". The engine receives a pre-set amount of fuel per cycle corresponding to each notch setting. Most modern diesel locomotive engines have eight such settings, plus an idling position. Notch 1 corresponds to slowest engine RPM, while Notch 8 corresponds to the highest. To get a train moving, the engine driver would let off the brakes, and put the throttle into Notch 1. Through this action, the driver also activates a notch-specific set of electronic relays, which hook the main generator to traction motors through a unique combination of contactors. Certain combinations of contactors result in higher voltage across traction motors while others result in lower voltage. Since traction motors produce more power at higher voltages, and vice versa, each notch setting corresponds to a certain power level. Diesel engines also require a battery to start the engine, and to provide electricity for lights, and controls when the engine is turned off.

2.5 Air Blower

Even though traction motors are very efficient machines, given their high power ratings, they still generate significant amounts of heat during operation, and thus require continuous cooling. This is accomplished through heavy duty air blowers, which derive their power from the diesel engine. Typical cooling systems involve an electrically powered fan, which blows air into cooling ducts. The air from these ducts is fed to the interior of traction motors, thereby cooling its armatures and then exits to the outside atmosphere. These blowers are also used to cool alternators. Sophisticated control systems are used to monitor temperatures of various rotating machines, and the volume of air supplied to individual systems is accordingly regulated.

2.6 Radiator and Radiator Fan

The diesel-electric locomotive has and additional cooling system for the engine block. The engine block is embedded with water lines, and water is pumped through these lines to cool it. The exiting water is cooled by passing it through a radiator, and its heat is blown away by a fan. This fan is driven by the diesel engine. The radiator and radiator fan are often located on the roof of the locomotive. A gearbox is used to transmit engine power to the fan and drive it.

2.7 Air Compressor

An air compressor is used to generate a constant supply of compressed air for the locomotive and train's brake system. This compressor is typically driven by the diesel engine directly.

Table 1 lists the specifications of a typical diesel-electric locomotive used by Indian Railways.

Table 1 Specifications of a diesel-electric locomotive used by Indian Railways

Manufacturers	Alco, DLW
Engine	Alco 251-B, 16 cylinder engine, 1900 kW, turbo-supercharged engine, 1000 rpm max, 400 rpm idle; bore × stroke: 228 mm × 266 mm; direct fuel injection, centrifugal pump cooling system: 2457 l/min, fan driven by eddy current clutch (64 kW at 1,000 rpm)
Governor	GE 17MG8/Woodwards 8574-650/Medha MEG 601
Transmission	Electric, with BHEL TG 10931 AZ generator (1,000 rpm, 770 V, 4520 A)
Traction motors	GE752 on original Alco models (302 kW), BHEL 4906 AZ (324 kW) and newer 4907 AZ
Axle load	18.5 long tons
Total weight	111 long tons
Bogies	Cast frame trimount bogies
Length over buffer beams	15.862 m
Distance between bogies	10.516 m

Source Wikipedia [8]

3 Sources of Noise in Diesel-Electric Locomotive

The noise from a diesel-electric locomotive is not attributable to a specific and single source. Rather, it comes from a large number of sources. And the nature of noise emanating from each source is characteristically different. Studies have shown that most of the noise in a diesel-electric locomotive is attributable to six different sources. These are diesel engine, generator, traction motors, air blowers for traction motors, cooling fan used in radiator, and structural vibration of the overall system. Looking at these sources individually can help in not only developing a richer understanding of overall noise, but also in crafting effective approaches for reducing diesel locomotive noise.

3.1 Engine Noise

The production of engine noise in itself is a very complex phenomenon. The noise from a diesel engine can be attributed to three principal sources. These are:

- *Combustion of gases*: This noise is attributable to changes in pressure inside the cylinder. These cyclic pressure fluctuations cause elastic deformations in the cylinder, which in turn cause the engine cylinder to radiate noise. The stiffness of the engine block has a significant influence on the magnitude of sound radiated outwards from the cylinder block. The higher the stiffness, the lesser is the magnitude of deformation of the engine-block and so is the amount of

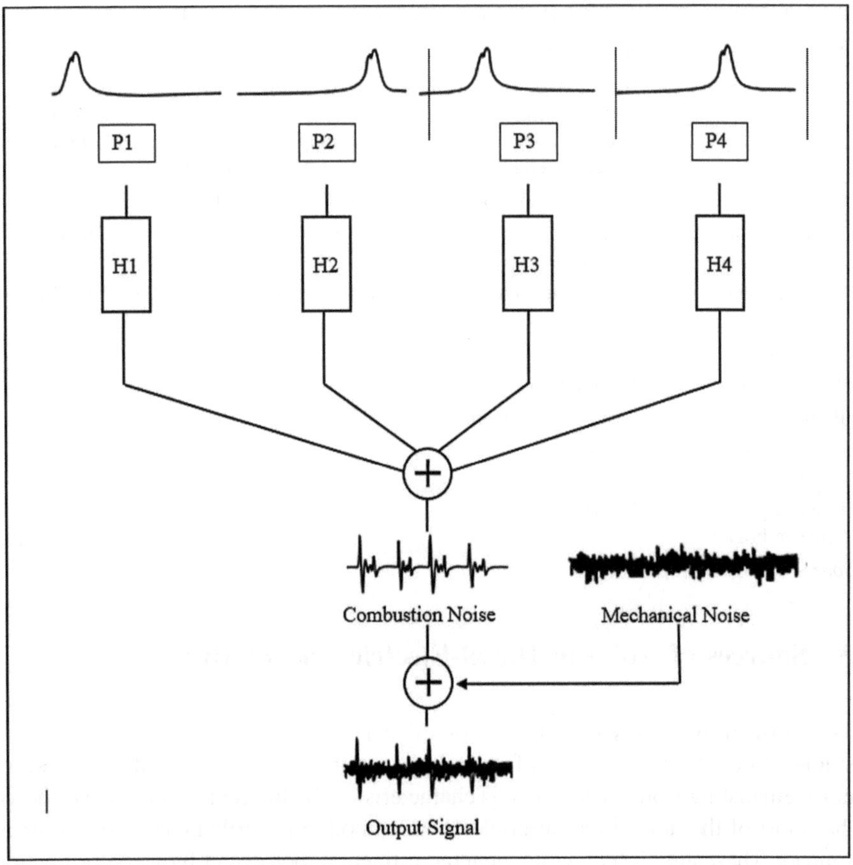

Fig. 2 Transmission of combustion noise

acoustic power radiated from the block. The acoustic transfer function between the inside of the cylinder and the outside is known as *structural attenuation filter* [9–13]. Thus for a four-cylinder system, there will be four such transfer functions, as shown in Fig. 2.

Here, four noise signals (P_1, P_2, P_3, and P_4) from four different cylinders, which are off from one another by a certain phase, get transmitted to the outside after getting attenuated by four different functions; H_1, H_2, H_3, H_4. These signals, post transmission to the outside, add up, and also get mixed with mechanical noise of the engine. This noise is finally sensed by the external microphone. For a single cylinder engine, the structural attenuation filter (or function), H, can be experimentally determined by placing a loudspeaker inside the cylinder, sealing the cylinder, and then computing H by measuring the SPL inside and outside the cylinder. While doing so, it should be ensured that the external microphone is placed at a distance sufficiently far enough from the cylinder so that the engine block's directivity does not affect the experimental data materially. This distance must be several times larger than the dimension of the cylinder block. A limitation of this method for measuring H is that it does not account for the change in cylinder volume during engine operation.

The character of combustion noise from a diesel locomotive differs from that of a regular automotive CI engine in two important ways. The first difference relates to its spectral content. The RPM of an automotive engine can vary between 500 and 5000 RPM, i.e. 8.33 and 83.3 RPS. In terms of frequency, this is a range corresponding to 3.33 octaves. For a diesel engine used for locomotive applications, the operating RPM range is narrower, and is typically between 269 and 904 RPM, i.e. 4.48 and 15.1[1] RPS. Thus the corresponding spectral bandwidth would be 1.75 octaves. Because of this, spectral content of combustion noise from these engines is narrower. Second, directivity effects for locomotive diesel engines are stronger given that these engines are large in size. Low frequency content dominates this noise, because the fundamental tone associated with this noise, which is a multiple of RPS and number of cylinders, does not usually exceed 250 Hz. If the number of significant overtones does not exceed 6, then most of the combustion noise energy is contained in frequencies less than 1500 Hz. However, higher frequency components can exist especially in presence of knocking. Such irregular self-ignition of fuel generates very large increments in cylinder pressure over extremely short durations. When such a pressure time-series signal is resolved in frequency domain, high frequency components also emerge. Noise attributable to such harsh and irregular auto-ignition of fuel is particularly unpleasant [14].

- *Mechanical components of engine*: This noise is attributable to mechanical operation of various components of the diesel engine system. Principal components contributing to this noise are oil pan, valve covers, cylinder-block sides, exhaust manifolds, rear and front sides of engine, intake manifolds, oil cooler, tappet plates, fuel pump, piston, etc. Amongst these, pistonic motion generates particularly unpleasant piston-slap noise. This noise is generated because the piston moves back and forth upon the wrist pin inside the bore of engine-cylinder. In process, it makes a tapping noise which gets transferred through the solid case, and the ensuing noise gets radiated through the surface. Jenkins [15] has tabulated the noise contribution from several of these components for a V-engine. Data from this works is reproduced in Table 2.
- *Exhaust gases*: Exhaust is a significant contributor to diesel engine noise. Compared to gasoline engines, exhaust noise in diesel engines is significantly stronger. This is due to a stronger "blow-down" event vis-à-vis petrol engines. Since the internal pressure in cylinder chamber is higher in diesel engines,

[1]For a multi-cylinder engine, as is the case for diesel locomotives, the upper and lower frequency limits should be multiplied by the number of cylinders in the engine. For a typical ALCO 251-B engine used by Indian Railways, the number of cylinders is 16. Assuming that the firing sequence is equally spaced, this would imply that fundamental frequency of combustion noise would be in the range of 36–121 Hz.

Table 2 Noise contributions from different components of a V-engine

Component	Noise in dB(A) at 3 ft distance
Oil pan and valve covers	99 each
Cylinder block sides	94
Exhaust manifold	92
Engine rear and front	91 each
Oil cooler	85
Tappet plates	85
Fuel pump	81

Source Jenkins [15]

exhaust gases "blow-down" the cylinder at relatively higher pressures and velocities. Thus exhaust noise levels for diesel engines, including the ones used in diesel locos, is higher vis-a-vis gasoline engines. The frequency content of this noise is dominated by the firing frequency. For a multi-cylinder four-stroke engine, this value is given by the following relation.

$$\text{Firing frequency} = (\text{RPM} \times \text{Number of cylinders})/120$$

However, as pressure waves in gases exiting through the exhaust are periodic though not harmonic, exhaust noise contains not only the firing frequency tone, but also its harmonics. Further, gases exiting the cylinder do not necessarily have nice planar and periodic wave fronts. There is turbulence as well. Thus, exhaust noise always has some broadband component.

In general, sound waves propagating along a pipe can be attenuated using either a dissipative or a reactive muffler. A dissipative muffler uses sound absorbing material to absorb acoustic energy present in exiting gases. These mufflers tend to perform well at higher frequencies, but are particularly inefficient at lesser frequencies. Reactive mufflers, reflect the sound waves back towards the source and prevent it to leave the pipe and reach outside. These mufflers are often designed through use of acoustic transmission line theory. These mufflers work either on the principle of a Helmholtz resonator, or an expansion chamber. The response of these mufflers is best at a particular frequency. In general such silencers are used to reduce low-frequency noise.

3.2 Cooling Fan Noise

Cooling fan is the principal source of noise in a diesel-electric locomotive. Cooling fans are usually mounted on the roof of the locomotive. Since overall diesel engine efficiency is only around 30%, most of the heat generated due to diesel combustion is wasted away. While some this heat is released to the environment through exhaust, a significant fraction is transmitted to the environment through the radiator.

Radiator cooling is carried out through forced convection driven by cooling fans. As the heat released to the outside through the radiator is very large, these fans have to push humongous amounts of air across radiator fins. For instance, the WDP-4 diesel-electric locomotive used by Indian Railways employs a 54-in. radiator fan which provides 77,500 CFM of air [16].

There are three sources of noise in these machines; the electric motor which drives them, the fan itself, and the ducts through which air is supplied. Noise generated by these fans has a broadband component due to vortex shedding from fan blades, and also a tonal component attributable to passage of blades against inlet and outlet openings. These fans can supply as much as 30–50 m^3 of air each second. Noise generated from fans is radiated to the surroundings through three paths. First, it may be radiated from the outlet or the inlet, if there is no duct attached to these ports. The second path constitutes the housing of the blower. Finally, sound may be also generated by vibrations induced in the adjoining structure due to mechanical coupling between the fan housing and the "base". Among these three transmission paths, it is relatively easy to arrest spread of noise due to structural coupling between fan housing and engine body.

The noise level due to a fan at some distance away is quantified in terms of sound pressure level L_p. Under STP conditions, the value of L_p in decibels (dB) at a given distance r, due to a sound-generating source having a sound power level of L_w (in dB) can be expressed as:

$$L_p = L_w + \mathrm{DI} - 20\log_{10}(r) + 10\log_{10}(e^{-mr}) - 10.9. \qquad (1)$$

Here, m is the energy attenuation coefficient, which is twice that of attenuation coefficient α, and DI is the directivity index of the noise source. There are three steps involved in calculating overall L_p attributable to fan noise. First, L_w, which is the total sound power level attributable to blower is calculated using the following relation [17].

$$L_w = L_w(\mathrm{B}) + 10\log_{10}(Q/Q_o) + 20\log_{10}(P/P_o) + \mathrm{B_T}. \qquad (2)$$

In this relation L_w(B) is basic sound power level corresponding to a particular fan type, and a particular octave band, Q is volume flow rate in ft^3/min, Q_o is reference flow rate which is 1 ft^3/min, P is rise in air pressure across the fan, P_o is reference pressure which is 248.8 Pa, and B_T is blade tone component. The value of B_T is zero at all frequencies except for that octave band in which blade pass frequency (f_b) lies. The value of f_b can be calculated by multiplying rotational speed of fan in RPS, and the number of blades on the fan. The values for L_w(B) and B_T have been tabulated by Graham in [18].

Next, power radiated by three different paths is computed using the following relations.

$$
\begin{aligned}
L_{\mathrm{w\_outlet}} &= L_{\mathrm{w}} - 3\,\mathrm{dB}\ \text{if outlet is open,} \\
L_{\mathrm{w\_inlet}} &= L_{\mathrm{w}} - 3\,\mathrm{dB}\ \text{if inlet is open, and} \\
L_{\mathrm{w\_housing}} &= L_{\mathrm{w}} - TL.
\end{aligned}
\tag{3}
$$

Here *TL* corresponds to transmission loss specific to an octave band, and it is specified by the blower manufacturer. It should be noted that radiated power level as calculated using Eq. 2 is specific to an octave band. Finally, corresponding to each octave band, total radiated power in decibels is computed using Eq. 3. This total radiated sound power level for each band is then used to compute sound pressure levels specific to different corresponding octave bands using Eq. 1. Finally, the overall sound pressure level is computed. It must be noted that there may be several other blowers units on the locomotive. If their contributions are significant those need to be calculated as well using the method explained here.

3.3 Noise Due to Structural Vibrations

The diesel-electric locomotive contains a large number of machines generating vibrations. Amongst these, the diesel engine is the principle source of these vibrations. Vibrations from these machines get transmitted to the locomotive body. These transmitted vibrations in turn excite large surfaces of the locomotive to vibrate and generate sound. The amplitude of such vibro-acoustic noise can become particularly large if the natural frequency of the excited portion of locomotive equals the excitation frequency. This is the condition for resonance. The amplitude of such vibro-acoustic noise depends on four parameters; amplitude of excitation frequency, area of the excited surface, proximity of the excitation frequency to the natural frequency of excited surface, and the amount of structural damping present in the material at the frequency of interest. If the velocity map of the excited surface is known, then it can be used to calculate volume velocity associated with the vibrating surface. This map may then be further used to compute sound pressure level at a specified distance.

There are no simple empirical or analytical relations to compute overall values of L_{p} for vibro-acoustic noise due to complexity and variety of noise generation phenomena, and also due to the presence of a very large number of uncontrolled variables. Thus, estimation of this noise has to be done on a case-to-case basis through a combination of detailed analytical, experimental, and statistical methods. Such an exercise is especially needed if structural vibrations constitute a significant source locomotive noise.

3.4 Traction Motor Blower Noise

Traction motor blowers also generate noise. However, these noise levels are significantly less intense vis-à-vis those produced by cooling fans. This is because the magnitude of flow rate provided by these blowers is typically an order of magnitude less than that provided by cooling fans. Thus, the contribution of volumetric flow rate to sound power level (L_w) as per Eq. 2 could be lesser for these blowers by as much as 10 dB.

3.5 Noise from Traction Motors

The noise from traction motors can be attributed to three different phenomena [19]. Aerodynamic effects due to passage of cooling air across windings, and blowing of fans across the motor are the first source of this noise. This noise is broadband as well as tonal in nature. Then there is noise due to magnetic effects. This noise has two components: load-dependent noise, and load-independent constant level noise. The latter component is generated due to interaction between magnetic flux produced by field coils and rotating magnetic parts of the motor. The former component of magnetic noise is generated due to time varying load-dependent current induced in rotor bars. This induced current creates a magnetic field around rotor bars, which attracts stator teeth in tangential and radial directions. These forces generate vibrations and noise in the motor.

Such effects are particularly strong in AC traction motors due to dimensional changes produced due to time-varying magnetic flux. Magnetic noise is highly tonal, as its frequency directly depends on frequency of the power line. Finally, traction motors are also noisy due to presence of mechanical sources including bearings, etc.

Over last several years, there have been significant reductions in noise produced by traction motors. Much of these reductions have been made possible through reductions in the aerodynamic component of motor noise. Presently, much of the traction motor noise is attributable to magnetic effects. This noise is often experienced as a "whining" sound. This is particularly true for trains driven by electric locomotives. In diesel locomotives, even though this noise is present, it gets masked by other types of noise, which have far higher sound pressure levels. The same is also true of noise from other electric motors used in diesel-electric locomotives.

3.6 Noise from Generators

The overall sound power level produced by generators, sans the drive, can be calculated using the following equation [20].

Table 3 Octave band specific corrections for generator noise

Octave band (Hz)	Central frequency (Hz)	Value to be deducted from L_w in dB
22–44	31.5	11
44–88	63	8
88–177	125	7
177–355	250	7
355–710	500	7
710–1420	1000	9
1420–2840	2000	11
2840–5680	4000	14

Data from Noise and Vibration Control for Mechanical Equipment, Washington DC, Technical Manual, TM 5-805-4 AFM 88-37 NAVFAC Dhl 3.10, Army, Airforce, and Navy (1983), Washington, DC

$$L_w = 84 + 10\log_{10} MW + 6.6\log_{10}\mathrm{RPM}. \tag{4}$$

Here, *MW* corresponds to megawatts of power produced by the generator. To obtain sound power levels in specific octave bands, corresponding values as shown in Table 3 should be subtracted from L_w.

4 Noise Reduction Strategies for Diesel-Electric Locomotives

In general, noise produced by a diesel-electric locomotive is attributable to a large number of sources present in the locomotive. Identifying and characterizing the contribution of every single source can consume significant amount of time and resources. Hence, it is important to have an estimate of the relative contribution of each noise source. Such analysis should form the basis for a diesel-electric locomotive noise reduction strategy.

Following such an approach will yield a relatively simple, effective, less expensive and less time consuming noise reduction strategy. As per Barron [21], there are four principal sources which when taken together generate almost 100% of total noise produced by a typical diesel-locomotive engine. These sources and their relative contributions are listed in Table 4.

Table 4 Noise contributions from a 3000 HP Diesel-Electric Locomotive

Noise source	Fraction of sound power emitted (%)
Engine exhaust	52
Cooling fan	41
Traction motor blower	6
Engine vibration	1

Data from Barron [17]

Additionally, there is also perceptible noise due to interaction between wheels and the track, and due to vibrations of structural parts of the locomotive. In another work [22], exhaust muffler, engine, fan, wheel-rail interaction, and electrical generator have been listed as top five sources of diesel-electric locomotive noise.

These data suggest that the best way to reduce noise from a diesel-electric locomotive is by reducing the noise from top two contributors; engine exhaust and cooling fan. Elimination of noise contributions from these sources could reduce overall noise reduction by as much as 10.5 dB. It is only after noise from principal sources has been significantly reduced, that addressing other sources will make sense. Thus an overall assessment of contributions from principal noise-sources constitutes the first element of a good noise reduction strategy for diesel locomotives.

The second important element of an effective noise reduction approach involves determination of spectral characteristics of the noise emanating from all principal sources. Such characteristics need to be measured on dB as well as dB(A) scales. This information will enable the acoustic engineer to help address following important questions.

- For a given noise source, how much noise is present in each octave band? The answer to this question will help the acoustic engineer select those noise reduction approaches which work well in specific bands. Certain noise reduction approaches work well for lower frequencies, while others are more suitable for high frequency noise.
- What is the ratio of tonal acoustic energy and broadband acoustic energy for noise attributable to a particular source? The answer to this question will enable the acoustic engineer to reduce tonal noise if needed. For instance, magnetic noise generated by traction motors has a high degree of tonality. Thus, noise mitigation measures such as Helmholtz resonators which perform very well at specific frequencies may be apt solutions for reducing such noise.
- How does A-weighting affect spectral characteristics of noise? Many regulatory requirements require noise, as measured on the dB(A) scale, to be less than a certain threshold. Such standards use the dB(A) scale because human beings perceive noise on this scale. The dB(A) scale assigns less weight to low frequency noise content, vis-à-vis that around 1000 Hz. The A-weighting curve is shown in Fig. 3.

Thus, resolving spectral characteristics of noise on A-weighted scale will help craft noise reduction approaches especially in situations when where noise reduction goals are set on the dB(A) scale.

Once these questions have been addressed, appropriate noise reduction methods must be identified for each important noise source. Figure 4 is a listing of these methods.

These noise reduction methods may be grouped as *passive* and *active* noise control methods. In contrast to the former methods of noise reduction, the latter group of noise control methods uses feedback or feed-forward electronic control mechanisms. Both these methods are described further.

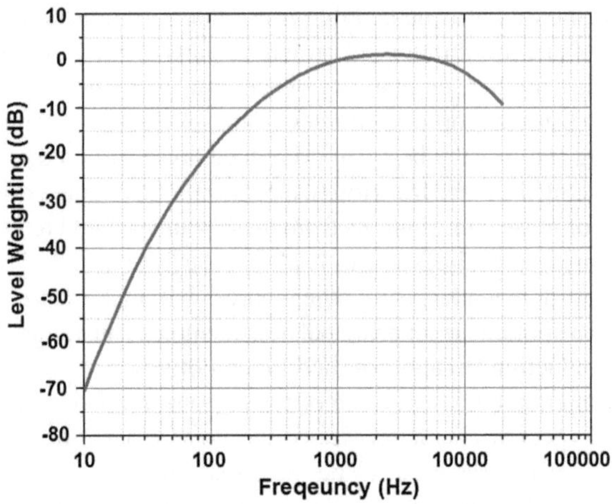

Fig. 3 A-weighting scale

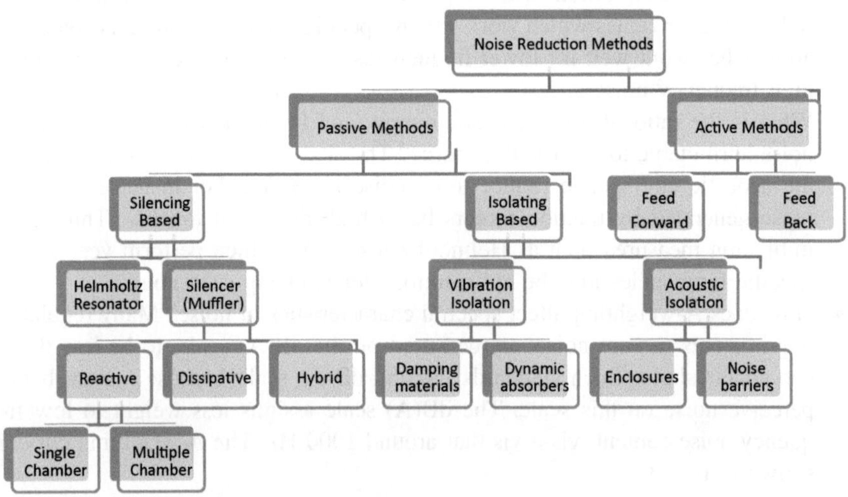

Fig. 4 Overview of different noise control methods

4.1 Passive Approaches to Noise Reduction

There are two types of passive methods used for noise reduction; "silencing" methods, and "isolating" methods. In methods belonging to the "silencing" category noise is either reflected back, or absorbed. Passive approaches, where noise gets reflected back into the system are termed *reactive*, while the latter method of noise reduction, where noise gets absorbed and gets converted into heat is designated as *dissipative*. Then there are noise reduction methods belonging to the "isolating" category. In these methods, propagation of air-borne or vibro-acoustic noise is physically arrested within a region through a combination of barriers, enclosures, and vibration isolators.

Silencing methods for noise reduction: Mufflers, also known as silencers, reduce noise using the silencing method. There are several types of mufflers, principal ones being side-branch mufflers, expansion type silencers, dissipative mufflers, and plenum chambers.

The *side-branch muffler* uses a Helmholtz resonator to reduce noise in a very limited frequency range. Here, the resonator is connected to the duct through which sound is transmitted. Figure 5 depicts two topologies for such a device. The resonator part of the muffler is designed to resonate at a specific frequency, f_o, as defined in Eq. 5.

$$f_o = (2\pi M_A C_A)^{-0.5} \tag{5}$$

Here, M_A and C_A correspond to resonator's acoustic mass and compliance of the resonator, respectively and are defined in [23]. If this resonance frequency matches that of the sound passing through the main duct, significant noise reduction is achieved primarily by bouncing of acoustical energy back to the source. Additionally, a small portion of acoustical energy gets dissipated inside the resonator due to its acoustic resistance. Since the resonator of such a muffler has only one resonance frequency, the working range of the muffler is restricted to a narrow bandwidth. However, if the resonator is replaced by a long tube closed at one end, then such a side-branch muffler will work at multiple frequencies. These frequencies are defined in Eq. 6.

$$f_o = (n - 0.5)c/(2L_c) \tag{6}$$

Here, n is a natural number, c is the speed of sound, and L_c is the equivalent length of the closed tube. In another configuration, the side branch resonator can be replaced by a short open tube, or an orifice. Design details for all these three flavors of side-branch muffler can be found in [24]. Such mufflers are cost effective

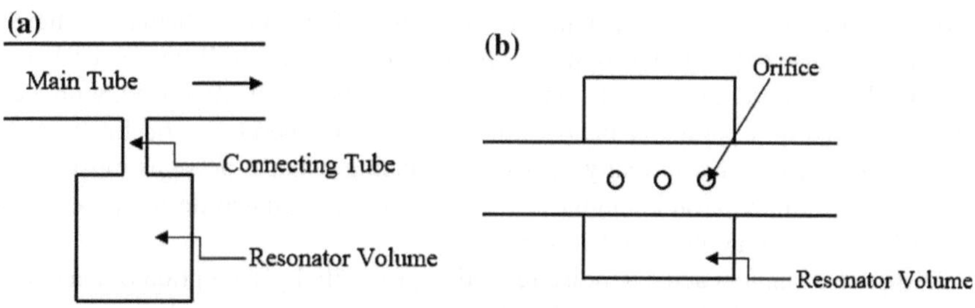

Fig. 5 Side-branch mufflers: **a** resonator volume connected by tube **b** resonator volume connected through orifices

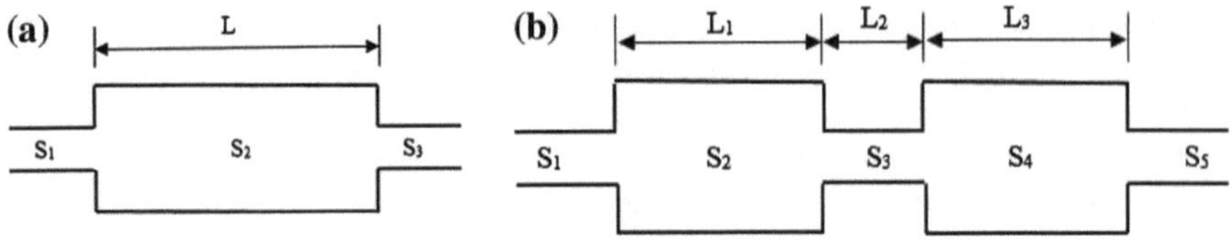

Fig. 6 Configurations for expansion chamber mufflers: **a** single expansion volume **b** double expansion volume

solutions for noise reduction if most of the noise is monotonic. Examples of such noise may include that from motors and engines running at constant rpm.

The *expansion chamber muffler* is also a reactive type muffler. These silencers have one or more expansion volumes connected in series with the duct through which sound is transmitted. Typical configurations for such mufflers are shown in Fig. 6.

If the ratio of cross-sectional areas of inlet and exit of the muffler is unity, then the relation for transmission loss (*TL*) for such mufflers is

$$TL = 1 + 0.25(m - 1/m)^2 \sin^2(kL). \tag{7}$$

In this relation, *m*, *k*, and *L* correspond to the ratio of cross-sectional area of muffler to that of the inlet pipe, the wave number, and the muffler's length, respectively. It is seen from Eq. 7, that the muffler is most effective when the term $\sin^2(kL)$ is unity. This corresponds to the condition $2f_o = (n - 0.5)c/L$, where *n* is an integer. For this condition, transmission loss across the muffler is maximum, and equals $1 + 0.25(m - 1/m)^2$. Thus, the performance of such mufflers peaks at specific frequencies. As one moves away from these optimal frequencies the performance of these mufflers reduces, initially very slowly, and later rapidly to eventually become zero for the condition $\sin(kL) = 0$. This is so because the derivative of *TL* with respect to *kL* is directly proportional to $\sin(2kL)$, and it equals zero when $\sin^2(kL)$ is unity. For this reason, these mufflers can operate over relatively larger bandwidths vis-à-vis side-branch mufflers. Further, transmission loss in a single-chamber muffler is a local maxima at several frequencies, while the side-branch muffler employing a Helmholtz resonator works best only at a single frequency. For these reasons, expansion chamber mufflers are particularly efficient at reducing noise content in a frequency band of modest width. When several such bandwidths exist, multiple expansion chamber mufflers may be deployed, with each chamber tuned to a particular bandwidth. In context of diesel-electric locomotive, these mufflers can come in really handy for reducing exhaust noise as most of the noise is concentrated around engine's firing frequency. Detailed design procedures for such mufflers may be found in [25].

Dissipative mufflers achieve noise reduction primarily by absorption of energy as sound passes through the silencer. Unlike reactive mufflers, these devices perform over wide bands. Transmission loss curves for these silencers do not exhibit sharp peaks and valleys as is the case for reactive mufflers. Hence, these mufflers are used to reduce wide-band noise. Sources of such noise would include fans noise, gas turbines, and leakages in acoustic enclosures. Even though these devices perform over wide bands, they do not work well at low frequencies. This is so, because commercially available sound absorbing materials do not perform well at low frequencies. Hence, the payoffs of using dissipative mufflers are available particularly at medium and high frequencies. In contrast, reactive mufflers can be appropriately sized to work at low frequencies. For these reasons most commercial mufflers have a *hybrid design*. These mufflers combine features of expansion chamber, dissipative, and side-branch designs.

Isolating methods for noise reduction: An alternative method to reduce noise is through isolation. Radiated noise can be prevented from further spreading by use of enclosures and barriers. In contrast vibro-acoustic noise can be contained by restricting the spread of vibrations to regions as close as possible to the source of vibrations itself.

Acoustic enclosures are one way of achieving acoustic isolation. These devices are particularly effective when large reductions in noise are required, and when the magnitude of direct sound field dominates that of the reverberant sound field. Reductions in noise levels in the range of 20–30 dB can be achieved if the source is completely enclosed. However, use of such method restricts venting of internally generated heat, and easy access to the machine. In such cases, partial enclosures may be used. Acoustic enclosures are considered *small*, if the bending wave of the enclosure wall is large relative to largest wall dimension, and if the wavelength of sound is large compared to largest internal dimension of the enclosure. In such enclosures the air space inside the enclosure and the walls are acoustically coupled, and there is little sound attenuation due to surface absorption and wall transmission effects. Performance of these enclosures strongly depends on stiffness of the surround walls [26]. Stiffer enclosures will transmit less sound and vice versa. When acoustic enclosures have to be *large*, they have to be designed differently. In such enclosures, some incident sound gets absorbed by the walls, and another fraction of it gets blocked by walls. To capitalize on these noise attenuation effects, a well-designed enclosure wall needs to be multiple layered; with the inner layer having large surface absorption coefficient, and outer layers having large values for transmission loss. For locomotive applications, enclosures may be a good option for containing noise from generators, and traction motors. They may also be particularly useful for preventing propagation of noise into driver's cabin. *Acoustic barriers* are also used to control noise. However, they are more effective outdoors, where reverberant noise field is non-existent. For indoor applications, especially in locomotives, acoustic barriers may have limited applicability because reverberant sound field's intensity is comparable to that of direct sound.

Vibration isolation and reduction is needed for containing vibro-acoustic noise. The magnitude of vibro-acoustic noise depends on four parameters; amplitude of excitation frequency, area of the excited surface, proximity of excitation frequency to the natural frequency of the excited surface, and the magnitude of structural damping present in the material at the frequency of interest. Amongst these four parameters, it is practically difficult to control the first two parameters. While the modification of the first parameter requires changes in the design of vibrating machinery, improvement of second parameter requires change in overall layout of the structure on which the machinery is mounted. However, it is relatively less difficult to modify other two parameters.

Vibroacoustic noise can be reduced in several ways. If the excitation frequency is very close to the natural resonance of the structure, then modifications to structural design may be carried out to increase the difference between resonance and excitation frequencies. This may be achieved by use of stiffening features, change of material, or change of material thickness. Such an approach works well when the excitation frequency is sufficiently low. For situations when this is not true, ensuring natural frequency of the structure to exceed excitation frequency can be a very expensive proposition.

Another way to reduce structural vibrations is to reduce overall transmissibility through introduction of vibration isolation materials between the structure and the vibrating component. These materials work well particularly for high frequency applications. Common examples of such materials are cork, felt, rubber, elastomers, metal springs, and fiber glass.

A third way for accomplishing the same goal is through change of the material of vibrating panels from metals to plastics. In general, metallic panels resonate at much larger amplitudes vis-à-vis plastic panels. This is because relative to plastics, metals have very poor damping characteristics. This method works for a fairly large frequency range.

A fourth way to reduce vibro-acoustic noise is by attaching a tuned-mass-damper on the vibrating machinery. If such a device is tuned to the excitation frequency of the machine, it will act as a sink for much of the excitation energy, and not much energy will flow into the adjoining structure. Detailed structural dynamic analysis has to be conducted to ensure the efficacy of such devices before their actual implementation. This approach works very well when most of the excitation energy is contained in a single frequency.

4.2 Active Approaches to Noise Reduction

Active noise control (ANC) is an attractive proposition to achieve significant noise reduction in a small package. ANC works by introducing an "anti-noise" signal through loudspeakers, to "kill" the original noise signal. Since the original signal may be changing with time in terms of frequency and amplitude, there is an electronic system controlled through a signal processing algorithm, to keep track of these changes, and accordingly generate "anti-noise" for cancelling it. This technology works really well at low frequencies. At higher frequencies, the ANC system is usually not fast enough, and thus it does not work up to expectations.

ANC systems either use feedforward, or feedback control systems. In the feedforward system, a coherent *reference* noise input is sensed, and its cancelling equivalent is fed downstream through a secondary source at an appropriate time. In the feedback method, the ANC system attempts to cancel the noise in absence of an "upstream" reference input signal. Feedforward methods may be further classified as broadband feedforward and narrowband feedforward. A very large number of algorithms have been developed for ANC systems [27]. ANC can be an excellent option to reduce noise from two principle sources in a locomotive engine; exhaust, and cooling fan. In one study [28], it has been demonstratively shown that ANC can reduce exhaust noise of a diesel-electric locomotive by as much as 15 dB.

5 Noise Regulations, Standards, and Methods

There are several standards available across the world, which regulate locomotive and train noise. Amongst these, perhaps the US standards governing railroad noise emissions are most comprehensive and well defined. The following table lists US based regulations governing railroad noise emissions which particularly related to locomotive noise.

Amongst these, 40 CFR Part 201, and 40 CFR Part 210 cover noise in line haul and yard operations. Agencies responsible for enforcing these regulations are Environmental Protection Agency (EPA), and Federal Railway Authority (FRA), respectively. Following parts of CFR 201 cover locomotive noise in particular (Table 5).

- 201.11: Standard for locomotive operation when it is stationary
- 201.12: Standard for locomotive operation when it is moving
- 201.16: Standard for locomotive load cell test stands
- 201.23: Details of acceptable test site, weather conditions, and ambient noise while conducting noise measurement for locomotives, trains, and locomotive load cell test stands at a 30 m distance
- 201.24: Methods for measuring noise from locomotive, trains, and locomotive load cell test stands at a 30 m distance.

These regulations are enforced through 49 CFR Part 210 and are implemented by the American Federal Railway Authority (FRA). Then there is noise in locomotive cabins, which driver and other operating personnel are exposed to. To

Table 5 List of regulations governing locomotive noise emissions in us railway systems

Code of Federal Regulations (CFR) Number	Nature of regulation
40 CFR Part 201	Noise emission standards for interstate rail carriers
49 CFR Part 210	Railway noise emission compliance regulations
49 CFR Part 227	Occupational noise exposure

Source Handbook of Railroad Noise Measurement, US Department of Transportation, October 2009

protect personnel from such excessive noise, there is CFR Part 227, which is essentially based on Occupational Safety and Health Administration's (OSHA) noise standard. There are other regulations and standards as well which cover locomotive noise attributable to horns, braking, wheels, etc. However, this text refers to only those regulations that relate to locomotive engine noise.

5.1 Measuring Locomotive Noise in Line-Haul Operations

As per [29], noise from a moving locomotive should not exceed 90 dB(A). This noise has to be measured at site with specifications as shown in Fig. 7.

Further, the following requirements must be observed while measuring pass-by locomotive noise.

- Microphone should be located at a 30 m distance from the center-line of the track.
- The track should be straight; such that it should not have a radius of curvature less than 873 m. In other words, it should not be curved with an angle exceeding 2°.
- Track must meet the specifications as mentioned in [29].
- There should be a "clear zone" as shown in Fig. 7. This zone should be free of large reflecting objects.
- Measurement over water, snow covered ground, and high grass should be avoided.

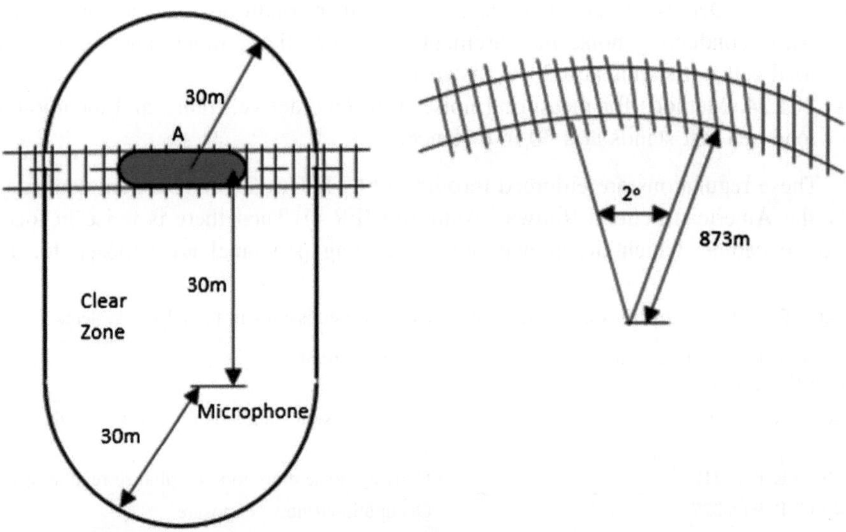

Fig. 7 Important details for measuring pass-by locomotive noise

- Measurement at sites which are closer than ¼ mile from a grade crossing or areas close to sources of electromagnetic interference should be avoided.

For measuring noise, A-frequency weighting with a "Fast" filter response, and "maximum sound pressure level" (L_{AFmx}) should be used. This should be done in the following way.

- Measure background noise level for 30 s, and its maximum value should be noted down.
- Next, conduct a pass-by measurement, and continuously measure sound level and note the maximum sound level. The pass by measurement should be initiated when the locomotive starts its approach. The measurement should continue till the locomotive is approximately 500 ft past the microphone line.
- Check for problems if any.
- Measure background noise level for 30 s, and its maximum value should be noted down.

For compliance purposes, noise levels only for a single pass-by event are required. However, for research and development purposes, multiple pass-by measurements are recommended to ensure increased accuracy and reliability.

5.2 *Locomotive Cab Noise*

49 CFR 227 sets noise exposure standards for railway employees spending significant amounts of their work hours in locomotive cabins. It is based on OSHA occupational noise standard, but with certain adaptations consistent with the railway environment. This regulation limits noise exposure of such personnel to 8-h time-weighted-average (TWA) of 90 dB(A). It also requires an effective "hearing conservation program[2]" for individuals exposed to noise in excess of 8-h TWA of 85 dB(A). The following table lists maximum noise exposure durations corresponding to different noise levels measured on dB(A) scale.

If noise in locomotive cabin exceeds the level as specified in Table 6, then the employee is required to use appropriate hearing protection device. For measuring noise exposure, a filter with "slow" response is recommended. Further, the regulation also permits exposure of noise in levels between 115 dB(A) and 120 dB(A) as long as total daily duration does not exceed 5 s.

[2]A "hearing conservation program" protects hearing of employees through training and use of hearing protection devices, and also monitors each employee's hearing ability through regular audiometric testing.

Table 6 Maximum exposure durations for different noise levels

Level of noise in dB(A)	Maximum exposure duration (h)
105	1
100	2
95	4
90	8
85	16
80	32

Source of data Table 14, Handbook of Railroad Noise Measurement, US Department of Transportation, October 2009

5.3 *Other Relevant Standards*

There are a very large number of additional standards, which may be useful in context of monitoring train noise. A partial list of these standards is provided below.

ANSI S1.1–1994. Acoustical terminology.

ANSI S1.13–1995 (R1999). Methods for the measurement of sound pressure levels in air.

ANSI S12.1–1983 (R1996). Guidelines for the preparation of standard procedures for the determination of noise emission from sources.

ANSI S12.2–1995 (R1999). Criteria for evaluating room noise.

ANSI S12.3–1985 (R1996). Statistical methods for determining and verifying stated noise emission values of machinery and equipment.

ANSI S12.4–1986 (R1993). Method for assessment of high-energy impulsive sounds with respect to residential communities.

ANSI S12.9/1–1988 (R1998). Quantities and procedures for description and measurement of environmental sound. Part 1.

ANSI S12.9/2–1992 (R1998). Quantities and procedures for description and measurement of outdoor environmental sound. Part 2: Measurement of long-term, wide-area sound.

ANSI S12.9/3–1993 (R1998). Quantities and procedures for description and measurement of environmental sound. Part 3: Short-term measurements with an observer present.

ANSI S12.9/4–1996. Quantities and procedures for description and measurement of environmental sound. Part 4: Noise assessment and prediction of long-term community response.

ANSI S12.18–1994 (R1999). Procedures for outdoor measurement of sound pressure level.

ASTM E596–96. Standard test method for laboratory measurement of noise reduction of sound-isolating enclosures.

ASTM E1014–84 (1995). Standard guide for measurement of outdoor a-weighted sound levels.

ASTM E1433–95. Standard guide for selection of standards on environmental acoustics.

ASTM E1574–98. Standard test method for measurement of sound in residential spaces.
ASTM E1686–96. Standard guide for selection of environmental noise measurements and criteria.
ASTM E1704–95. Standard guide for specifying acoustical performance of sound-isolating enclosures.
ASTM E1779–96a. Standard guide for preparing a measurement plan for conducting outdoor sound measurements.
AS1217.1–1985. Acoustics: determination of sound power levels of noise sources: guidelines for the use of basic standards and for the preparation of noise codes.
AS2377–1980. Methods for the measurement of airborne sound from rail-bound vehicles.
AS2900.7–1986. Quantities and units of acoustics.
IEC 61063–1991. Acoustics: measurement of airborne noise emitted by steam turbines and driven machinery.
IEC 61260–1995. Electroacoustics: octave-band and fractional-octave-band filters.
ISO 31–7–1992 Quantities and units—Part 7: Acoustics.
ISO 140/4–1998 Acoustics: measurement of sound insulation in buildings and of building elements. Part 4: Field measurements of airborne sound insulation between rooms.
ISO 226–1987. Acoustics: normal equal-loudness level contours.
ISO 266–1997. Acoustics: preferred frequencies.
ISO 230/5–2000. Test code for machine tools. Part 5: Determination of the noise emission.
ISO 362–1998. Acoustics: measurement of noise emitted by accelerating road vehicles: Engineering method.
ISO 1680–1999. Acoustics: test code for the measurement of airborne noise emitted by rotating electrical machines.
ISO 1683–1983. Acoustics: preferred reference quantities for acoustic levels.
ISO 1996/1–1982. Acoustics: description and measurement of environmental noise. Part 1: Basic quantities and procedures.
ISO 1996/2–1987. Acoustics: description and measurement of environmental noise. Part 2: Acquisition of data pertinent to land use.
ISO 1996/3–1987. Acoustics: description and measurement of environmental noise. Part 3: Application to noise limits.
ISO 1999–1990. Acoustics: determination of occupational noise exposure and estimation of noise-induced hearing impairment.
ISO 2017–1982. Vibration and shock: isolators. Procedure for specifying characteristics.
ISO 2151–1972. Measurement of airborne noise emitted by compressor/prime mover-units intended for outdoor use.
ISO 2204–1979. Acoustics: guide to international Standards on the measurement of airborne acoustical noise and evaluation of its effects on human beings.
ISO 3095–1975. Acoustics: measurement of noise emitted by railbound vehicles.
ISO 3381–1976. Acoustics: measurement of noise inside railbound vehicles.

ISO 4869–1–1990. Acoustics: hearing protectors: Part 1: Subjective method for the measurement of sound attenuation.
ISO 4869–2–1994. Acoustics: hearing protectors: Part 2: Estimation of effective A-weighted sound pressure levels when hearing protectors are worn.
ISO 4871–1996. Acoustics: declaration and verification of noise emission values of machinery and equipment.
ISO 7188–1994 Acoustics: measurement of noise emitted by passenger cars under conditions representative of urban driving.
ISO 7235–1991. Acoustics: measurement procedures for dueled silencers. Insertion loss, flow noise and total pressure loss.
ISO/TR 7849–1987. Acoustics: estimation of airborne noise emitted by machinery using vibration measurement.
ISO 11546/1–1995. Acoustics: determination of sound insulation performance of enclosures. Part 1: Measurements under laboratory conditions (for declaration purposes).
ISO 11546/2–1995. Acoustics: determination of sound insulation performance of enclosures. Part 2: Measurements in situ (for acceptance and verification purposes).
ISO 11688/1–1995. Acoustics: recommended practice for the design of low-noise machinery and equipment. Part 1: Planning.
ISO 11688/2–1998. Acoustics: recommended practice for the design of low-noise machinery and equipment. Part 2: Introduction into physics of low-noise design (at draft stage).
ISO 11689–1996. Acoustics: procedure for the comparison of noise-emission data for machinery and equipment.
ISO 11957–1996. Acoustics: determination of sound insulation performance of cabins: laboratory and in situ measurements.
ISO 13332–2000. Reciprocating internal combustion engines: test code for the measurement of structure-borne noise emitted from high-speed and medium-speed reciprocating internal combustion engines measured at the engine feet.
ISO 14163–1998. Acoustics: guidelines for noise control by silencers.
ISO 15667–2000. Acoustics: guidelines for noise control by enclosures and cabins.
CEN/TR 15874:2009. Railway applications—Noise emission—Road test of standard for rail roughness measurement EN 15610:2009.
CEN/TR 16891:2016. Railway applications—Acoustics—Measurement method for combined roughness, track decay rates and transfer functions.
CEN/TS 16272-5:2014. Railway applications—Track—Noise barriers and related devices acting on airborne sound propagation—Test method for determining the acoustic performance—Part 5: Intrinsic characteristics—In situ values of sound reflection under direct sound field conditions.
CEN/TS 16272-7:2015. Railway applications—Track—Noise barriers and related devices acting on airborne sound propagation—Test method for determining the acoustic performance—Part 7: Extrinsic characteristics—In situ values of insertion loss.

EN 15461:2008 + A1:2010. Railway applications—Noise emission—Characterization of the dynamic properties of track sections for pass by noise measurements.
EN 15610:2009. Railway applications—Noise emission—Rail roughness measurement related to rolling noise generation.
EN 15892:2011. Railway applications—Noise Emission—Measurement of noise inside driver's cabs.
EN 16272-1:2012. Railway applications—Track—Noise barriers and related devices acting on airborne sound propagation—Test method for determining the acoustic performance—Part 1: Intrinsic characteristics—Sound absorption in the laboratory under diffuse sound field conditions.
EN 16272-2:2012. Railway applications—Track—Noise barriers and related devices acting on airborne sound propagation—Test method for determining the acoustic performance—Part 2: Intrinsic characteristics—Airborne sound insulation in the laboratory under diffuse sound field conditions.
EN 16272-3-1:2012. Railway applications—Track—Noise barriers and related devices acting on airborne sound propagation—Test method for determining the acoustic performance—Part 3-1: Normalized railway noise spectrum and single number ratings for diffuse field applications.
EN 16272-3-2:2014. Railway applications—Track—Noise barriers and related devices acting on airborne sound propagation—Test method for determining the acoustic performance—Part 3-2: Normalized railway noise spectrum and single number ratings for direct field applications.
EN 16272-4:2016. Railway applications—Track—Noise barriers and related devices acting on airborne sound propagation—Test method for determining the acoustic performance—Part 4: Intrinsic characteristics—In situ values of sound diffraction under direct sound field.
EN 16272-6:2014. Railway applications—Track—Noise barriers and related devices acting on airborne sound propagation—Test method for determining the acoustic performance—Part 6: Intrinsic characteristics—In situ values of airborne sound insulation under direct sound field conditions.
EN 16286-2:2013. Railway applications—Gangway systems between vehicles—Part 2: Acoustic measurements.
EN 16727-2-2:2016. Railway applications—Track—Noise barriers and related devices acting on airborne sound propagation—Non-acoustic performance—Part 2-2: Mechanical performance under dynamic loadings caused by passing trains—Calculation method.
EN ISO 11690-1:1996. Acoustics—Recommended practice for the design of low-noise workplaces containing machinery—Part 1: Noise control strategies (ISO 11690-1:1996).
EN ISO 11690-2:1996. Acoustics—Recommended practice for the design of low-noise workplaces containing machinery—Part 2: Noise control measures (ISO 11690-2:1996).
EN ISO 14163:1998. Acoustics—Guidelines for noise control by silencers (ISO 14163:1998).

EN ISO 15667:2000. Acoustics—Guidelines for noise control by enclosures and cabins (ISO 15667:2000).
EN ISO 1680:2013. Acoustics—Test code for the measurement of airborne noise emitted by rotating electrical machines (ISO 1680:2013).
EN ISO 3095:2013. Acoustics—Railway applications—Measurement of noise emitted by rail bound vehicles (ISO 3095:2013).
EN ISO 3381:2011. Railway applications—Acoustics—Measurement of noise inside rail bound vehicles (ISO 3381:2005).
FprEN 16727-3. Railway applications—Track—Noise barriers and related devices acting on airborne sound propagation—Non-acoustic performance—Part 3: General safety and environmental requirements.
prEN 15610 rev. Railway application—Acoustics—Rail and wheel roughness measurement related to rolling noise generation.
prEN 16727-1. Railway applications—Track—Noise barriers and related devices acting on airborne sound propagation—Non-acoustic performance—Part 1: Mechanical performance under static loadings—Calculation and test methods.
prEN 16727-2-1. Railway applications—Track—Noise barriers and related devices acting on airborne sound propagation—Non-acoustic performance—Part 2-1: Mechanical performance under dynamic loadings due to passing trains—Resistance to fatigue.
prEN 16951-1. Railway applications—Track—Noise barriers and related devices acting on airborne sound propagation—Procedures for assessing long term performance—Part 1: Acoustic characteristics.
prEN 16951-2. Railway applications—Track—Noise barriers and related devices acting on airborne sound propagation—Procedures for assessing long term performance—Part 2: Non-acoustic characteristics.
prEN ISO 14163 rev. Acoustics—Guidelines for noise control by silencers.

References

1. SCI Verker (2012) Electric locomotive market primed for growth. Int Railway J 2012. http://www.railjournal.com/index.php/locomotives/electric-locomotive-market-primed-for-growth.html
2. Kevin S (2013) Diesel locomotive market growing despite weak European demand. Int Railway J. http://www.railjournal.com/index.php/locomotives/diesel-locomotive-market-growing-despite-weak-european-demand.html
3. Kevin S (2013) Diesel locomotive market growing despite weak European demand. Int Railway J. http://www.railjournal.com/index.php/locomotives/diesel-locomotive-market-growing-despite-weak-european-demand.html
4. https://www.google.co.in/search?q=cost+of+an+average+diesel+locomotive&rlz=1C1CHBD_enIN709IN709&oq=cost+of+an+average+diesel+locomotive&aqs=chrome..69i57.6749j0j4&sourceid=chrome&ie=UTF-8
5. Matthew T (2013) A comparison of diesel and electric locomotive noise emissions from coal terminal rail loop and spur line. In: Proceedings of acoustics—Victor Harbor, Australia, Australian Acoustical Society

6. https://en.wikipedia.org/wiki/Diesel_locomotive
7. https://en.wikipedia.org/wiki/Traction_motor
8. https://en.wikipedia.org/wiki/Indian_locomotive_class_WDM-2#Technical_specifications. 5B2.5D
9. Renard C, Polac L, Pascal JC, Sahraoui S (2004) Extraction of vibration sources in diesel engines. In: Proceedings of the 11th international congress on sound and vibration, St. Petersbourg, Russia
10. Boustany AR, Gautier F, Wang S (2006) Source separation in diesel engines with the cyclic Wiener filter. In: Proceedings of Euronoise, Tampere, Finland
11. El Badaoui M, Daniere J, Guillet F, Serviere C (2005) Separation of combustion noise and piston-slap in diesel engine—Part I: separation of combustion noise and piston-slap in diesel engine by cyclic Wiener filtering. Mechanical Syst Signal Process 19:1209–1217
12. Pruvost L, Leclere Q, Parizet E (2006) Diesel combustion noise extraction using Wiener filtering. Optimisation and validation of the method. In: Proceedings of the SIA's automobile comfort conference, Le Mans, France
13. Pruvost L, Leclere Q, Parizet E (2007) Improvement of the spectrofilter—separation of coherent sources overlapping in time and frequency domains. In: Proceedings of the 19th international conference on acoustics, Madrid, Spain
14. Pruvost L, Leclere Q, Parizet E (2009) Diesel engine combustion and mechanical noise separation using an improved spectrofilter. Mech Syst Signal Process 23:2072–2087
15. Jenkins JH (1975) Analysis and treatment of diesel-engine noise. J Sound Vibr 43:293–304
16. http://www.daulatram.com/cool1.htm
17. Barron RF (2003) Industrial noise control and acoustics. Marcek Dekker, Inc., NY, pp 164–167
18. Graham JB (1972) How to estimate fan noise. Sound Vibr 24–27
19. Le Besneraisa-b J, Lanfranchic V, Hecquetb M, Récorbeta S, Sapenaa J, Brochetb P (2008) Characterization of audible magnetic noise emitted by traction motors in railway rolling stock. In: Proceedings of Internoise 2008
20. Bies DA, Hansen CH Engineering noise control—theory and practice, 3rd edn. Spon Press, p. 554
21. Barron RF (2003) Industrial noise control and acoustics. Marcek Dekker, Inc., NY, p. 214
22. Wyle Laboratories for US Environmental Protection Agency (1971) Transportation noise and noise from equipment powered by internal combustion engines, a report prepared. Washington DC, pp 135–137
23. Beranek LL (1993) Acoustics. Acoustical Society of America
24. Barron RF (2003) Industrial noise control and acoustics. Marcek Dekker, Inc., NY, pp 350–368
25. Munjal MM (2014) Acoustics of ducts and mufflers. Wiley (2014)
26. Ver IL Reduction of noise by acoustic enclosures. In: Isolation of mechanical vibration, impact, and noise, (1). ASME, New York, pp 192–220
27. Kuo Sen M, Morgan Dennis R (1999) Active noise control: a tutorial review. Proc IEEE 87:943–973
28. Cotana F, Rossi F (2002) Active noise control technique for diesel train locomotor exhaust noise abatement. J Acoust Soc Am
29. Handbook of Railroad Noise Measurement, US Department of Transportation, 2009, 29

Part III
Alternate Fuels for Locomotive Traction

Biodiesel as an Alternate Fuel for Diesel Traction on Indian Railways

Anirudh Gautam, Ravindra Nath Misra and Avinash Kumar Agarwal

Abstract Indian Railways (IR) have a fleet of 5000 diesel locomotives and this population is growing at a rate of 200 locomotives per year. Diesel expenditure on IR is ~Rs. 20,000 Crore per year (US$ 4 bn/Y), which is ~10% of the turnover of the IR. This is the main motivation for both, reducing this expenditure as well as replacing diesel with domestically produced alternate fuels. With this objective, Engine Development Directorate (EDD) of Research Designs and Standards Orgaisation (RDSO) has carried out engine studies using different indigenously produced biodiesels on two locomotive engines on engine test benches. Details of these engine studies and scope for future research on use of biodiesel for rail road traction are discussed in this chapter.

1 Introduction

Diesel locomotive engine is the most efficient energy conversion device with very high thermal efficiencies in the range of 37–40% for rail road traction. In addition, diesel locomotives reduce the burden on electric traction, since there is a shortage of ~25–30% electricity in India. Ministry of New and Renewable Energy (MNRE) of the Government of India has framed "National Biofuels Policy" for India, which is aimed to ensure availability of a minimum level of biofuels in the market, in order to meet the demand in near future.

The following definitions of biofuels are given in the National Biofuels Policy document:

A. Gautam (✉) · R.N. Misra
Ministry of Railway, New Delhi, India
e-mail: gautam.anirudh@gmail.com

A.K. Agarwal
Engine Research Laboratory, Department of Mechanical Engineering,
Indian Institute of Technology Kanpur, Kanpur 208016, India

A.K. Agarwal et al. (eds.), *Locomotives and Rail Road Transportation*,
DOI 10.1007/978-981-10-3788-7_5

- 'biofuels' are liquid or gaseous fuels produced from biomass resources and used in place of, or in addition to, diesel, petrol or other fossil fuels for transport, stationary, portable and other applications.
- 'biomass' resources are the biodegradable fraction of products, wastes and residues from agriculture, forestry and related industries as well as biodegradable fraction of industrial and municipal waste.

The scope of the National Biofuels Policy encompasses bio-ethanol, biodiesel and other biofuels:

- Bio-ethanol: Ethanol produced from biomass such as sugar containing materials, like sugar cane, sugar beet, sweet sorghum, etc.; starch containing materials such as corn, cassava, algae etc.; and cellulosic materials such as bagasse, wood waste, agricultural and forestry residues etc.
- Biodiesel: A methyl or ethyl ester of fatty acids produced from vegetable oils (edible and non-edible) or animal fat having properties similar to diesel.
- Other biofuels: Bio-methanol, bio-synthetic fuels etc. [1].

Biodiesel is quite similar to mineral diesel. The biggest advantage of biodiesel is that it is renewable and mitigates pollution emanating from the engine exhaust. IR purchases ~2.7 billion liters of diesel at a cost of ~Rs. 20,000 Crore annually. Even 5% replacement of mineral diesel by biodiesel will lead to a saving of ~Rs. 1,000 Crore annually, which can be used for providing financial support for increasing train speeds on IR tracks, make up for the shortage of electricity, provide assistance to agriculture based industries and reduce country's dependence on foreign exchange. An indicative target of 20% blending of biofuels, both for biodiesel and bio-ethanol, by 2017 is proposed. Blending levels prescribed for biodiesel are intended to be recommendatory in near-term. The blending level of bio-ethanol has already been made mandatory from October 2008 onwards and will continue to be mandatory leading up to the indicative target [1].

2 Diesel Locomotive Biodiesel Eco-System

Figure 1 illustrates the eco-system for use of biodiesel in diesel locomotives. First and foremost is the availability of raw materials. Abbott estimated that the role of biomass as an alternate fuel is about 0.008% of the total solar energy availability which is roughly about 7 TW (Table 1) [2]. This is ~20% of the total energy consumption of the world (~35 TW), therefore a significant amount of world's energy requirement can be met by biofuels [2]. Abbott further stated that the final solution for the world's energy needs will require a basket of alternate fuels. Biodiesel will be an important component in this basket.

Wild crops, especially cultivated energy crops and inexpensive imported fats can be the raw materials for producing biodiesel. States like Chattisgarh, Madhya Pradesh, Uttar Pradesh, Karnataka and Orissa have abundant availability of wild oil

Fig. 1 Diesel locomotive biodiesel eco-system

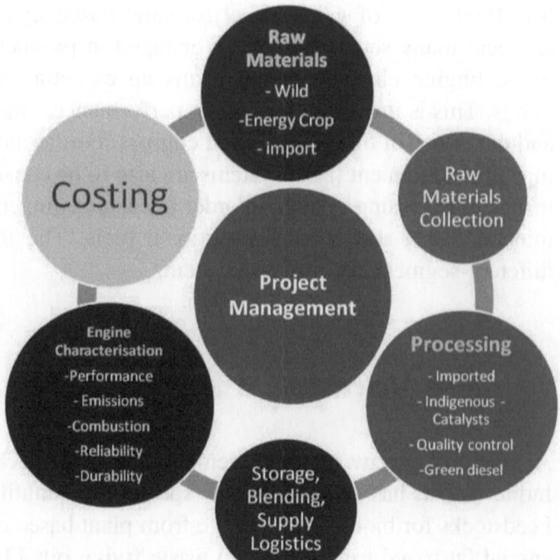

Table 1 Different energy sources and their proportion vis-a-vis solar energy [2]

Energy source	Max power (TW)	% of total solar energy
Total surface solar	85,000	100
Desert solar	7,650	9
Ocean thermal	100	0.12
Wind	75	0.08
Geothermal	44	0.05
River hydroelectric	7	0.008
Biomass	7	0.008
Open ocean wave	7	0.008
Tidal wave	4	0.003
Coastal wave	3	0.003

seeds, which need an organised eco-system for collection and delivery to the biodiesel production plants. This is an area, which needs participation of NGOs, Government agencies as well as private organizations. Processing of biodiesel from different raw materials using various types of chemical processes is well established. Both imported and indigenous technologies are available; however there is preference for established imported technologies in setting up biodiesel plants due to their superior organisation, structure and quality control in these plants. This is an area where Indian manufacturers need to work further to compete in international market. Storage, blending and supply logistics form an important part of eco-system since biodiesel and other biofuels are organically derived and have limited shelf life. Production of green-diesel (organic based hydrocarbons) is a fast emerging area and many setups have been engaged in production of green-diesel from biomass. Engine characterisation forms an essential part of the overall scheme of things. This is important for engine performance, emissions, combustion, reliability and durability of biodiesel fuelled engines. Furthermore, warranty issues of the fuel injection equipment (FIE) systems are also to be considered and resolved. Final part is the entire costing system, in order to ensure competitiveness of biodiesel vis-à-vis mineral diesel and other conventional fuels. The following sub-sections discuss different segments of this eco-system.

2.1 Raw Materials

Availability of raw materials remains the biggest challenge for biofuel policy of India. India has 300 different species of unutilized/underutilized oil seeds. Feedstocks for biodiesel can range from plant based edible and non-edible oils, and animal fat based triglycerides to waste frying oils [3]. Present availability of these oil seeds is roughly 25 million tons per year. Pongamia, Mahua, Cottonseed and Jatropha are some of the locally available raw materials, which can be used to produce biodiesel. Use of locally available feedstocks for biodiesel production has a positive influence on agriculture, energy security and strengthening of local and national economy. Availability of these seeds in India is shown in Table 2. Yearly production of biodiesel from these locally available non-edible oils can be in the range of 500–600 million liters, which is sufficient for IR to introduce at least a 20% biodiesel blend (B20) on its entire fleet of diesel locomotives.

Another route is development of energy crops similar to Malaysia and Indonesia. India can continue to import palm styrene and PFAD from these countries till it becomes self-reliant in the cultivation and collection of oil seeds. Future of bioenergy is estimated to include large proportions of ligno-cellulosic, biomass and Algae based green-diesels. Adequate R&D in these areas is therefore essential.

Table 2 Locally available non-edible oil seeds for production of biodiesel in India [4]

Non-edible seed/oil	Availability/year
Cottonseed oil	0.3 million tons
Waste fish oil	0.27 million tons
Pongamia seeds	0.6 million tons
Mahua seeds	0.18 million tons

2.2 *Processing and Different Pathways*

Production of biofuels including biodiesel has well established processes and pathways. Imported technologies like the one from Lurgi, Germany is quite popular in India. Indian Institute of Chemical Technoloy (IICT) and Indian Institute of Petroleum (IIP) have developed their own fully functional technologies with indigenous catalysts, where in all types of lipids and high Free Fatty Acids (FFA) oils can be converted to biodiesel. Some of the oil refineries have started direct blending of bio-based raw materials in petroleum stream and are calling the product as green-diesel because the product is a bio-derived hydrocarbon, which is quite similar to mineral hydrocarbons, unlike esters. Hydro-treatment is another route, which has been established to convert biomass into green biofuels. Breakdown of ligno-cellulosic materials has been achieved through various processes but economically viable systems/techniques are yet to be developed. Some R&D organisations are working towards creating commercially viable Algae based biofuels however substantial work is required in this area. Figure 2 shows various pathways, which can be used to convert biomaterials into fuels.

Fermentation and dehydration of sugars (cane, sugar beet, etc.) and starches to ethanol is a well-established process. 5% ethanol blended with gasoline (E5) is already commercially available in India, although there are issues related to supply and demand. Conversion of lignin, cellulose and hemi-cellulose to green cellulose

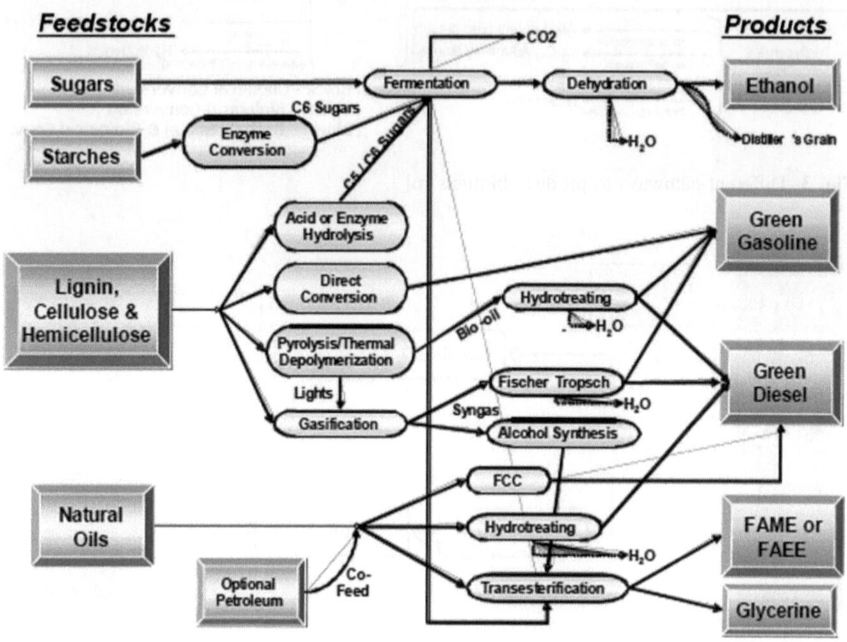

Fig. 2 Various pathways to convert biomaterials into biofuels [5]

or green-diesel through direct conversion, thermal depolymerisation and/or gasification is being tried on a large-scale by some oil companies such as Virent, USA. Availability of bio-hydrogen or renewable hydrogen is the main challenge in this area. Conversion of natural oils and heavy mineral oils through fluid catalytic conversion (FCC) is a well established technique. Hydro-treatment is an area under evolution and transesterification to FAME or FAEE and glycerine is well proven. More details on different pathways to produce biofuels are given in Fig. 3.

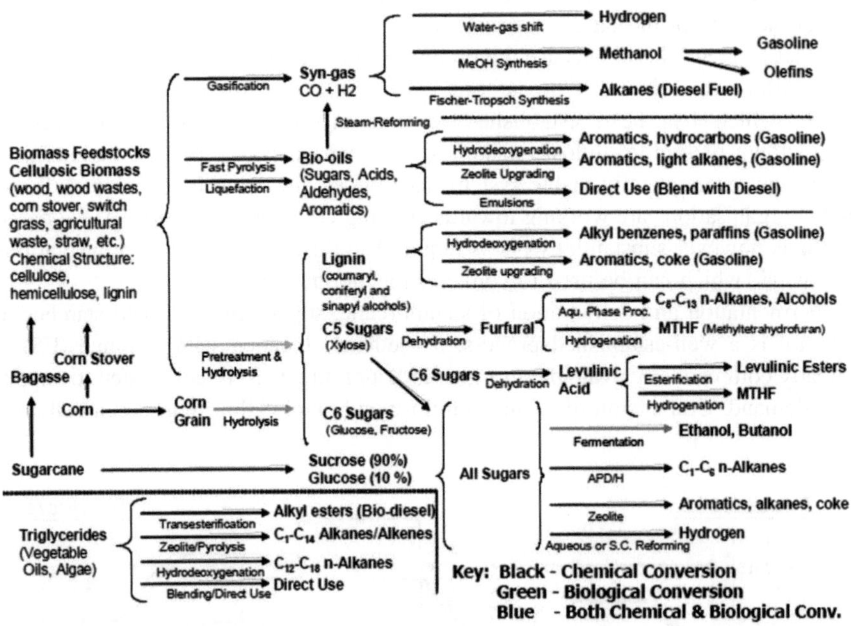

Fig. 3 Different pathways to produce biofuels [6]

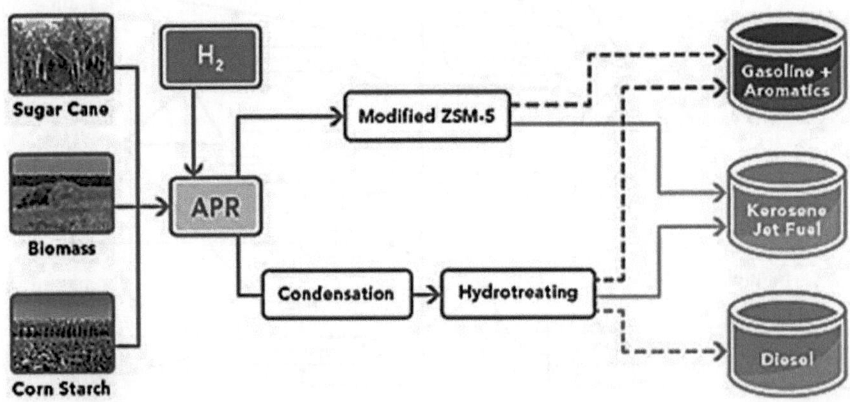

Fig. 4 Schemcatic of biomass conversion plant (Virent, USA) [7]

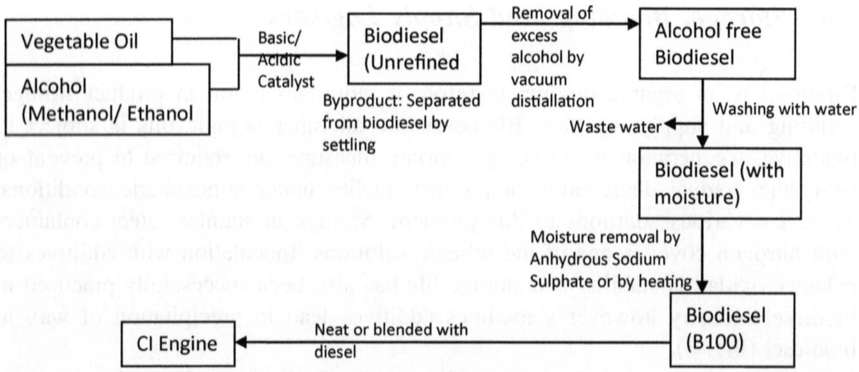

Fig. 5 Schematic of traditional biodiesel production process [3]

Schematic of an existing biomass conversion plant into diesel, kerosene and gasoline is given in Fig. 4.

Schematic of traditional biodiesel production using transesterification process is shown in Fig. 5.

2.3 *Quality Control*

The National Biodiesel Accreditation Program is a cooperative and voluntary program for the accreditation of producers and marketeers of biodiesel called BQ-9000® [8]. The program is a unique combination of ASTM standards for biodiesel (ASTM D6751) and a quality systems program that includes storage, sampling, testing, blending, shipping, distribution and fuel management practices. BQ-9000® is open to any biodiesel manufacturer, marketeer or distributor of biodiesel and biodiesel blends in the United States and Canada [8]. India has also issued its own biodiesel (B100) standard [BIS-15607] and Europe has brought out EN-14241 standard for biodiesel (B100). On examination of all these standards, it emerges that there are 22 parameters, which are required to be within stipulated limits. EN-14241 has additional requirements and these are included in the 22 parameters mentioned above. More than meeting these standards, it is the process control, which is emphasized in BQ-9000. Indian Industry must adopt BQ-9000 both for the plant construction and their operation, if it is to compete internationally. In India, IOC R&D facility at Faridabad was the first laboratory to establish complete biodiesel parameter testing facility. Later IICT and BPCL also set up testing facilities for testing complete biodiesel specifications. RDSO has also setup a facility for biodiesel testing at Lucknow, which is partially operational. In summary, qualification of production plant capabilities is required and process audit for production units is essential, so that they can produce biodiesel, which meets all the prevailing specifications.

2.4 Storage, Blending, and Supply Logistics

Biodiesel is an organic product therefore it requires caution in product storage, handling and supply logistics. Biodiesel like all other organic oils is subject to oxidative degeneration therefore appropriate measures are required to prevent or delay the gradual degradation in product quality under atmospheric conditions. There are various solutions to this problem. Storage in stainless steel containers with nitrogen cover is one of the reliable solutions. Inoculation with additives to enhance oxidation stability and storage life has also been successfully practised in biodiesel industry however sometimes additives lead to precipitation of wax in biodiesel (Fig. 6).

In 2010, Society of Automotive Engineers (SAE) formed a sub-committee (TC-7: Fuels committee) to explore various aspects of use of biodiesel in railroad applications. The sub-committee was setup against the backdrop of Energy Independence and Security Act of 2007 of USA and additional mandates by state governments in USA. ASTM-D975 diesel standards allow up to 5% blending of biodiesel with mineral diesel. Automotive companies had addressed many issues

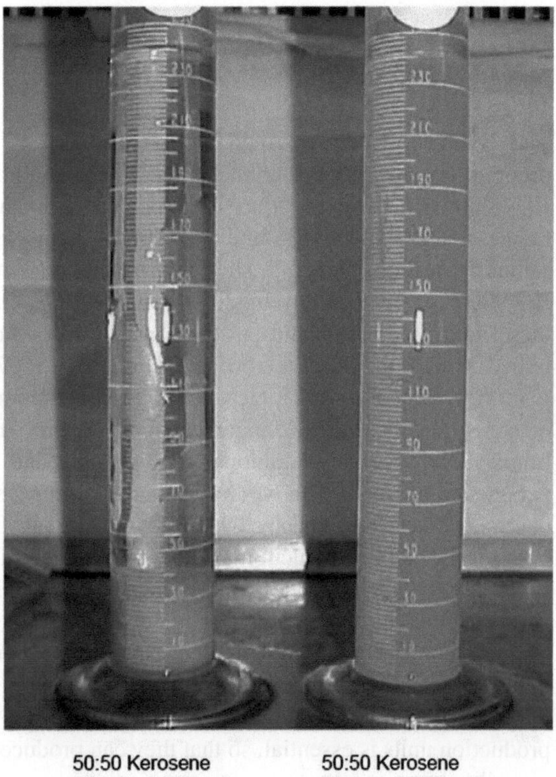

Fig. 6 Wax settling with different additives [9]

related to use of biodiesel and blends in Automotive engines in last several years, however Railroads/locomotive builders have not been as active and have expressed several concerns. Fundamental mandate of the SAE sub-committee was to identify few of these concerns related to use of biodiesel in diesel locomotives. Another important consideration was that the fuel suppliers were required to provide biodiesel and their viewpoints and concerns also needs to be taken into account since they were also an important part of the supply-chain. SAE sub-committee suggested the following areas for consideration:

- Emission (NO_x, PM and smoke): NO_x bump, variance issues in the past, voluntary agreements from fleet-average NO_x caps in California, claims, risk of taking results from small engines and applying them to larger, slower speed locomotive engines.
- Engine durability
- Fuel handling and material compatibility: Elastomers, filters, lines
- Cold temperature operability
- Fuel stability: Storage, oxidation, and thermal stability
- Corrosion: Steel tank and equipment
- Long-term durability: Pre-combustion (fuel tank to injection pump), combustion, post-combustion (blow-by gas)
- Lubrication oil dilution/stability, and
- Biological contamination.

Out of these topics, three main categories were selected and the working groups discussed them threadbare. These included emission, engine durability, and fuel handling and material compatibility. Following issues were listed regarding fuel handling, storage and material compatibility.

Technical considerations

Biodiesel is compatible with steel, aluminum and fiber-glass tanks. B20 and lower blends are compatible with most hoses and gaskets. Higher blends than B20 may require different hoses or gaskets than commonly used with mineral diesel. Biodiesel blends in any proportion with mineral diesel and does not separate, however stirring may be required for adquate mixing in some cases. Pure biodiesel is a good cleaning agent and can clean entire fuel systems upon initial use. Cleaning effect is very low in case of B20, and generally not noticeable in case of B5 blends.

Water and microbial contamination

B20 and lower blends of biodiesel hold similar quantities of water as mineral diesel. Saturation point of B20 and lower biodiesel blends are 50–250 ppm. Pure biodiesel-water saturation point is slightly higher (1000–1800 ppm) than that of mineral diesel. Water in excess of saturation point will separate and get desposited in the bottom of tanks similar to mineral diesel. This allows biodiesel to be treated similar to diesel (i.e. monitor for water, remove if needed). Microbial contamination found in biodiesel and blends can be treated in the same way as with mineral diesel.

Other renewable fuels

Biodiesel is the first renewable fuel that can be commercialized to replace diesel. "Biomass based diesel", "Non-ester renewable diesel", "Green-diesel", "Hydrogenated Vegetable Oils" and "Pyrolysis Oils" are some of the other renewable fuels. ASTM is in the process of determining, which of these new fuels are covered by ASTM D975, and which will need newer specifications or controls, especially for minor components or non-HC compounds.

Biodiesel Quality Throughout Supply Chain

BQ-9000 has recommended the following for maintaining biodiesel quality throughout the supply chain [8].

Before Receiving the Product

- Specify the respective ASTM certification for the fuel being produced.
- Consider purchasing from a BQ-9000 certified marketeer.
- Use of a fuel stabilizer is recommended for storing fuel beyond six months.
- Discuss your fuel's application and cold flow expectations with your supplier.
- Ensure storage tanks are clean and dry.

Upon Receipt of the Product

- Request a Bill of Landing or Certificate of Analysis upon fuel delivery.
- Perform ASTM haze rating evaluation to identify water and sediments.
- Retain two quart samples of your delivery for analysis, should problems arise.

Blending and Testing Strategies

- Be sure ASTM D975 diesel, ASTM D396 heating oil and ASTM D6751 biodiesel meet their specifications independently, prior to blending.
- Blends must be clear in appearance and free from water and sediments.
- B100 should be stored and blended at 10 °C above the cloud point.
- Adhere to BQ-9000 quality protocols.

Storage and Dispenser Strategies

- Fuel tanks should be kept as full as possible to minimize water contamination.
- Monitor hoses, fills, vents and gaskets for leaks.
- Check tanks monthly for water by obtaining a tank bottom sample, remove water, when required to prevent microbial contamination.
- Handle all blends of biodiesel as you would do for any diesel or heating oil.

If Problems Occur

- Contact your supplier, report your problem and ask for their assistance.
- Share your retained samples with your supplier for analysis.
- If an acceptable resolution is not reached, contact the National Biodiesel Troubleshooting Hotline (800-929-3437) and Participating organizations of USA in biodiesel program.

Similar approach is needed for biodiesel program in India to succeed and flourish, particularly for locomotive applications of IR.

ASTM has adopted a stability requirement for B100 and biodiesel blends. ASTM specification D6751 for B100 requires a minimum Rancimat induction time of 3 h. ASTM specification D975 (diesel specifications) allow up to 5% biodiesel, which meets the ASTM D6751 specifications. Based on data, B5 is assumed to provide a minimum 6 h Rancimat induction time. ASTM D7467 specifications for B6 to B20 blends require a minimum of 6 h Rancimat induction time.

2.5 *Biodiesel Aging and Stability*

Three aging situations can be considered for biodiesel and blends [9].

1. In storage and handling: Applies to B100 and blends. A maximum storage period of 6 months is recommended for B100 currently.
2. In vehicle fuel tank: Areas of concern are the recirculation at low fuel level, (only for biodiesel blends).
3. Aging (reacting) in high temperature engine fuel system: It will include deposit formation from unstable or pre-aged fuel and applies only to biodiesel blends.

Aging in Storage as per ASTM D4625 consists of:

a. Quiescent aging at 43 °C for 12 weeks
b. One week equivalent to one month for petroleum fuel (4 × acceleration factor)
c. Measurement of total insoluble plus other fuel properties.

Vehicle tank aging as per ASTM D4625 has the following steps:

a. Quiescent aging at 80 °C for 1 week.
b. With ullage (vacant container space: the amount or volume by which a container, especially one for liquids, is short of being full), purge to ensure that test is not oxygen limited.
c. Measurement of total insoluble and other fuel properties.

High temperature fuel system aging (reacting): ASTM D6468: Thermal stability test on un-aged blend

- Perform accelerated stability tests on B100 samples covering the range available in the country.
- Select few B100 samples covering the range of stability and feedstocks for:
  - 12 week storage tests.
  - Blending with different diesels to produce numerous B5 and B20 blend combinations (ULSD, LSD, wide range of aromatic content).
  - Perform accelerated stability tests on the B5 and B20 blends, including high temperature test, simulating conditions in an engine fuel injection system.

- Select few B5 and few B20 blends for:
  - 12 week storage test.
  - One week test simulated fuel tank aging.
- Interpret results to address study objectives.

SAE TC7 subcommittee on biofuels enumerated the following about the stability of biodiesel and its tests [9].

Accelerated Stability Tests: These are laboratory tests to find out if a sample is stable in a short time. The sample is oxidized under severe or "accelerated" conditions. There are two main approaches for these tests.

(a) Induction period (Ranciment/Oil Stability Index (OSI)) determines the time till oxidation starts. Induction period is measured as time to initial formation of acids by oxidative degradation at 110 °C in presence of air. This test is relevant to storage stability.
(b) Deposit Formation test (ASTM D2274): In this test, amount of gum or deposit formed is determined after oxidation for a fixed time. Deposits are measured after 16 h at 90 °C in oxygen. This test is relevant to engine compatibility.

Figure 7 shows the test results of B100 long-term storage tests as per ASTM D4625. These tests were performed in a vessel, which was partially filled by biodiesel. B100 long-term storage tests were carried out for 12 weeks at 43 °C with vessel, which was open to air. Induction period was found to deteriorate to less than 3 h in less than 4 weeks except for highly stable additized samples. Stable samples showed little or no peroxide formation. As a next step, impact of B100 stability on B5 stability was studied by conducting accelerated tests. The results are shown in Fig. 8.

B100 stability appears to be an excellent predictor of blend stability, an IP (induction period) of 3 h for B100 ensures low deposits and long induction period for blend. B5 blend's long term storage tests were carried out as per ASTM D4625. The results are shown in Fig. 9.

B5 long-term storage tests were carried out at a temperature of 43 °C with vessel open to air. The only B5 that experienced an increase in acid value (AV) or insoluble was B100 with IP less than 3 h. Other samples did not degrade, since there was no increase in peroxides for stable B100. It was concluded that 3 h IP adequately protects B5 blends under these conditions. B5 blends vehicle fuel tank simulation tests were carried out for 1 week at a temperature of 80 °C with ullage purge intended to simulate aging in vehicle tank. Test results are shown in Fig. 10. B5 blends from B100 with >3 h IP showed little increase in AV or insoluble. 3 h IP adequately protects B5 blends under these conditions. However, being a new test, its meaning is uncertain and more such tests need to be carried out to have stable conclusions.

B5 was subjected to thermal stability tests. Test results are shown in Fig. 11. It is observed that B5 was thermally stable if produced from oxidatively stable B100 (IP > 3 h).

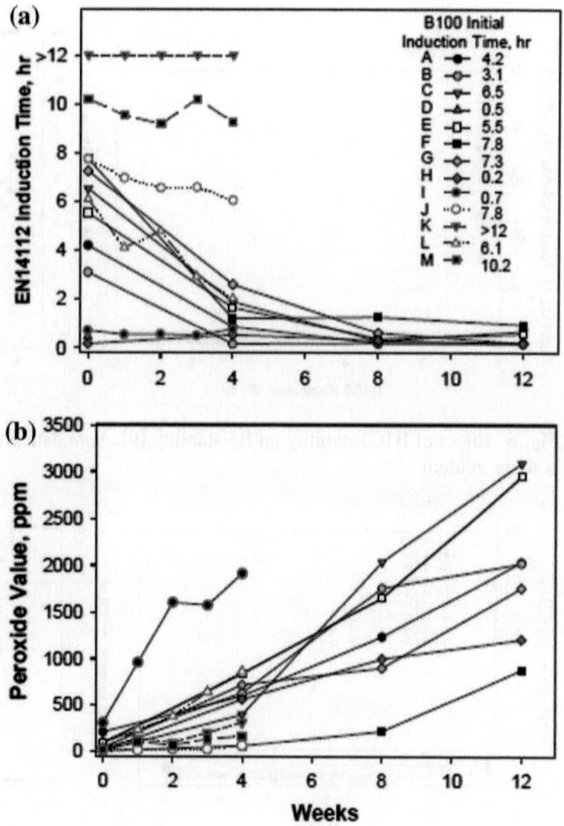

Fig. 7 B100 long-term storage test results as per ASTM D4625 [9]

Impact of B100 stability on B20 stability was studied with accelerated tests. Results are shown in Fig. 12. B100 stability appears to be an excellent predictor of blend stability, 3 h IP ensured low deposits and 6 h Rancimat in the biodiesel blend.

Long term storage test results for B20 as per ASTM D4625 are shown in Fig. 12. Tests were carried out for 12 weeks at 43 °C, with vessel open to air. No oxidation impacts were observed after 4 weeks. At 8 weeks, samples with < 3 h IP were found to degrade. Other samples with stable B100 didn't show any degradation, i.e. no increase in peroxides for stable B100. It was concluded that 3 h IP adequately protected B20 under these conditions.

Results of B20 vehicle fuel tank simulation studies are shown in Fig. 13. Fuel was stored for 1 week at 80 °C with ullage purge intended to simulate aging in vehicle tank. Test was not found to discriminate between stable and unstable blends for acid value or insolubles formation. Diesel produced upto 1 mg/100 ml total insoluble in this test.

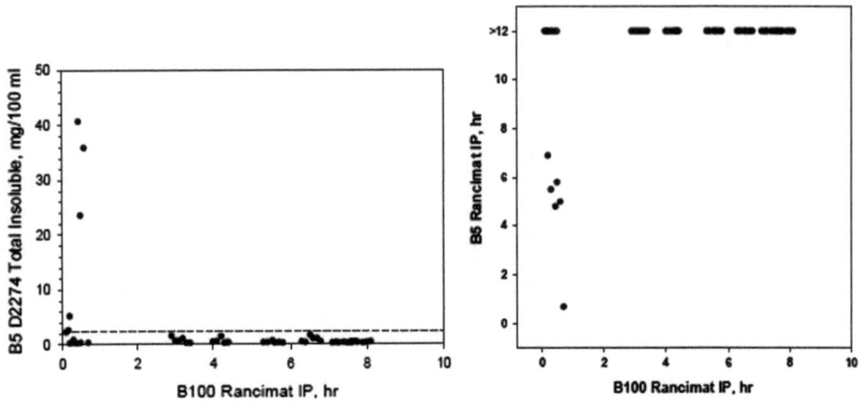

Fig. 8 Effect of B100 stability on B5 stability [9]. *Note* data points artificially spread out of points is more evident

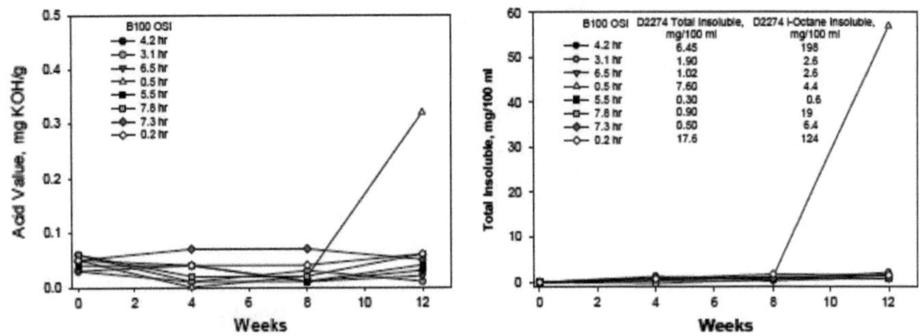

Fig. 9 B5 long-term storage tests [9]

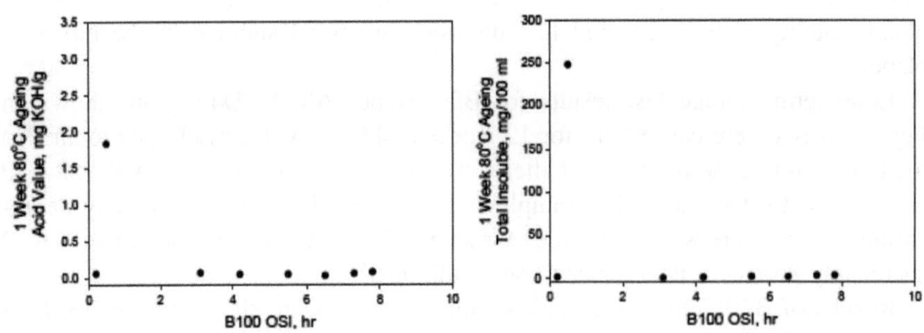

Fig. 10 Vehicle fuel tank simulation tests for B5 [9]

High temperature stability test results of B20 are shown in Fig. 14. B20 was thermally stable, if produced from oxidatively stable B100 with induction period of > 3 h (Fig. 15).

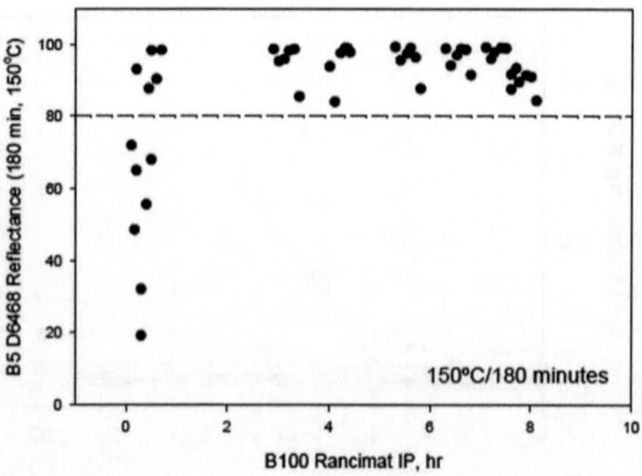

Fig. 11 Thermal stability tests for B5 [9]

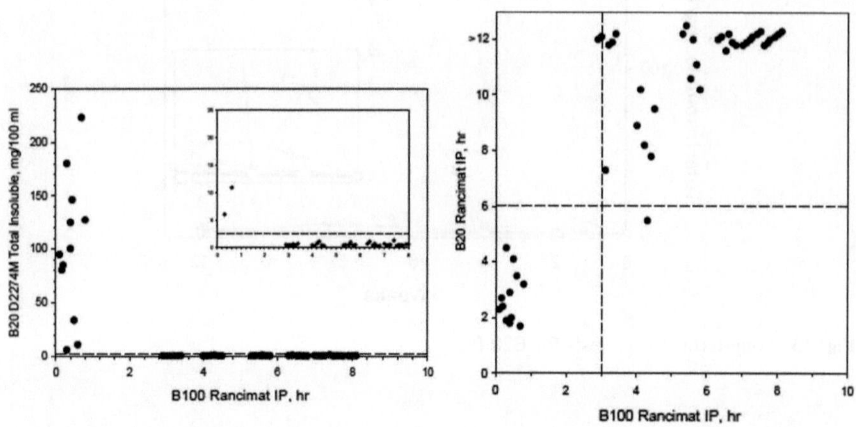

Fig. 12 Fuel stability tests of B20, dependence on B100 stability [9]. *Note* data points artificially spread out so number of points is more evident

Based on these experiments, it can be concluded that blend stability is dominated by B100 stability. It is independent of diesel stability, aromatic, or sulfur content. Most B100 samples begin to oxidize in storage, therefore use of synthetic anti-oxidants and limited storage time is recommended. A 3 h IP for B100 appears to be adequate to ensure stability of B5. For higher biodiesel blends, measurement of blend stability would provide added protection to the user. IP may need to be longer than 3 h at point of production to ensure stability at the point of blending. Hindered phenolic anti-oxidants prevent degradation of B100 and blends. Above conclusions apply to the specific test methods employed.

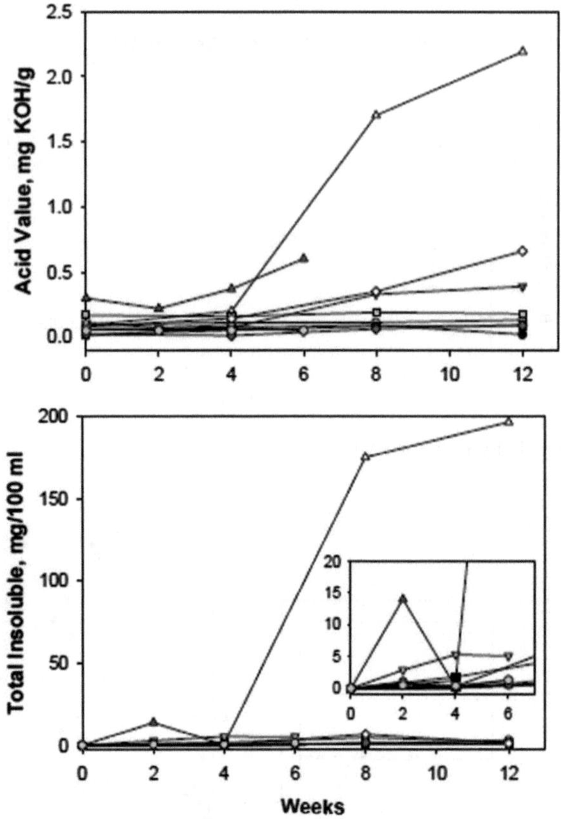

Fig. 13 Long-term storage tests for B20 [9]

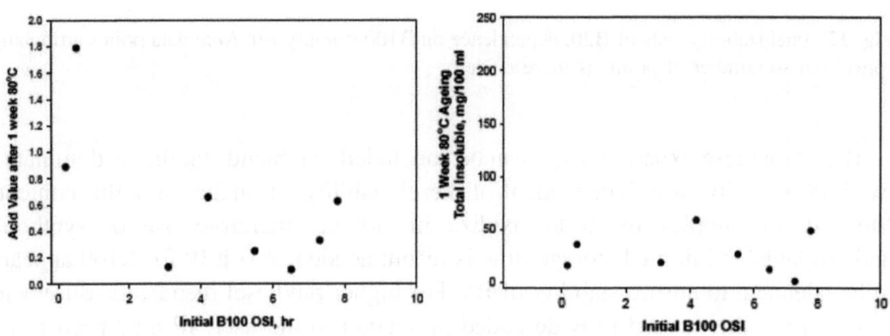

Fig. 14 B20 vehicle fuel tank simulation results [9]

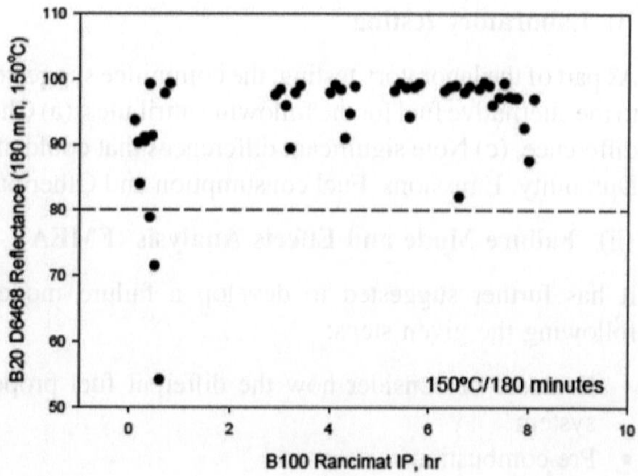

Fig. 15 B20 high temperature stability test results [9]

3 Engine Characterization

3.1 OEM Biodiesel Blend Approvals

Following blends of biodiesel were approved by various engine/automotive manufacturers [10]:

B5 approved—Audi, BMW, Mercedes, Volkswagen, Detroit Diesel, Freightliner, Isuzu, Mack, Volvo, Peterbilt, Sterling, Thomas built Buses, and Kubota. Honda, Nissan and Toyota are new entrants to US diesel market which are contemplating biodiesel blend approvals.

B20 supported—Ford, New Holland Agriculture, General Motors, Dodge, Case IHi Agriculture, Navistar International, Cummins, John Deere, Caterpillar, HDT, Ferris, Toro, Perkins Engines, ARCTIC CAT, Fairbanks Morse Engine (producing ALCO engines in USA), Yanmar, TomCar.

B100 approved—Following companies approve biodiesel usage upto B100: Case IH (approximately 50% of models), New Holland, Fairbanks Morse, Tomcar.

3.2 Suggested Process to Evaluate the Impact of Biodiesel in Locomotives [9]

The procedure to evaluate locomotive systems if alternative fuels are used, was debated. A multi-step process was suggested. The committee recommended following tests namely, (i) Analytical laboratory testing/evaluation of all fuels, (ii) Component endurance testing (iii) Engine performance testing (iv) Locomotive performance testing and (v) Multi-locomotive performance and endurance testing, in order to carry out Field performance testing and Endurance field evaluation study of impact on reliability and durability.

(i) **Laboratory testing**

As part of the laboratory testing, the committee suggested a comparison of current fuel to the alternative fuel for the following attributes, (a) Chemical difference, (b) Physical difference, (c) Note significant differences that could affect Performance, Reliability, Durability, Emissions, Fuel consumption and Other parameters.

(ii) **Failure Mode and Effects Analysis (FMEA)**

It has further suggested to develop a failure mode effect analysis (FMEA) by following the given steps:

- Evaluate or consider how the different fuel properties might impact complete system
- Pre-combustion
  - Fuel tank
  - Volumetric energy content (fuel tank volume and trip distance)
  - Suction pump (rust and solvency)
  - Lower pressure fuel lines (compatibility)
  - Fuel filters (compatibility and life)
  - Fuel heaters (copper on older fuel heaters)
  - High pressure pumps
- Combustion
  - Needle lift (lift timing changes because of pressure build up differences)
  - Injection spray (droplet size, penetration, rate, etc.)
  - Ignition delay (changed?)
  - Kinetic or premixed burn stage (changes?)
  - Diffusive burning (changes?)
  - Diffusive and end of burning period (extended?)
  - Ring groove deposit changes (increase?)
  - Cylinder liner (fuel impingement on wall?)
  - Fuel to power conversion (energy content differences)
- Post-combustion
  - Exhaust

    Temperature
    Emissions (NO_x etc.)
    Turbo speed changes?
  - Blow-by gases

    Crankcase oil useful life
    Base number depletion

Oxidation
Corrosion of lead (Pb, organic acids)

– Engine condition changes (piston under crown deposits, other changes to condition?)

(iii) **Fuel (biodiesel)**

It was suggested to establish the B100 source, B100 quality, Blending accuracy, BX biodiesel, Seasonal issues, Stability, Storage, Emissions, Energy content, Availability of supply and Cost. It was suggested to have a very detailed protocol to evaluate use of biodiesel on diesel locomotive fleet of American Railroads. Some of the elements of the process suggested have already been done by RDSO and IR.

3.3 *Biodiesel Effects on Engine Emissions*

The sub-committee discussed the effect of biodiesel on engine emissions [9]. He reviewed the tests conducted by US EPA on biodiesel fuelled engines.

3.3.1 US EPA Review of Biodiesel Emissions

Engine dynamometer test results were compiled by EPA, primarily on-highway engines. Engine models through 1998 and data taken from published studies conducted by others for different purposes were used. There was a 12% decrease in PM and CO on an average for B20 and 20% decrease in HC. Concurrently there were decreasing emissions of toxic compounds, e.g. carbonyl, PAH as observed from the limited data (Fig. 16).

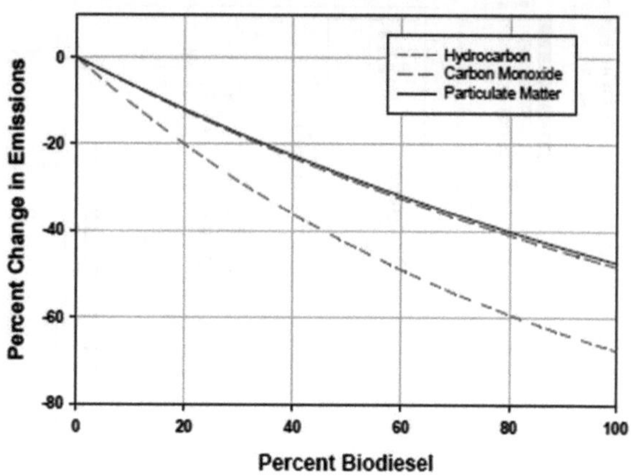

Fig. 16 Changes in emissions for different blends of biodiesel [9]

3.3.2 Results of Tests Conducted for Biodiesel Fuelled Engines by Other Researchers

Toxic emissions were lower for biodiesel as shown in Fig. 17. These include the C_1–C_{12} hydrocarbon emissions, total aldehydes, relative Ozone forming potential, selected Poly Aromatic Hydrocarbons (PAHs) and selected nitro-PAHs. US Environmental Protection Agency reviewed the NO_x emissions from biodiesel fuelled engines. They studied NO_x emissions from soy-based biodiesel fuelled engines. It was found that change in NO_x for B20 blend ranges from −7% to 7%, however B20 average change in NO_x was +2% and no appreciable change in NO_x was observed with B5 blends. Their study included 43 engines, in which 75% engine were pre-1994 and 95% engines were pre-1997. Many B20 tests showed decreasing NO_x emissions. Some factors that can affected the NO_x emissions from biodiesel fuelled engines were the feedstock, engine technology and the driving cycle.

EPA also characterized the heavy duty 2-stroke diesel engines. These 2-stroke engines were characterized by higher power to weight ratio. 1980s-early 1990s on-highway engines showed a larger NO_x effect for biodiesel. PM was found to sometimes increase for biodiesel fuelled 2-stroke engines. 2-stroke engines also showed higher PM soluble organic fraction (SOF). Engines with higher SOF emissions typically showed little impact of B20 on total PM emissions (Fig. 18).

Figure 18 shows the effect of SOF % in total PM for no. 2 diesel vis-a-vis percent change in PM in B20. It is seen that at lower SOF in total PM, the change in PM in B20 was negative and as the this value increased to higher side, the percent change in PM did not change much and remained almost constant with exception of one case being neglected. This indicated that the amount of SOF in the total PM did not affect the total change in PM in B20.

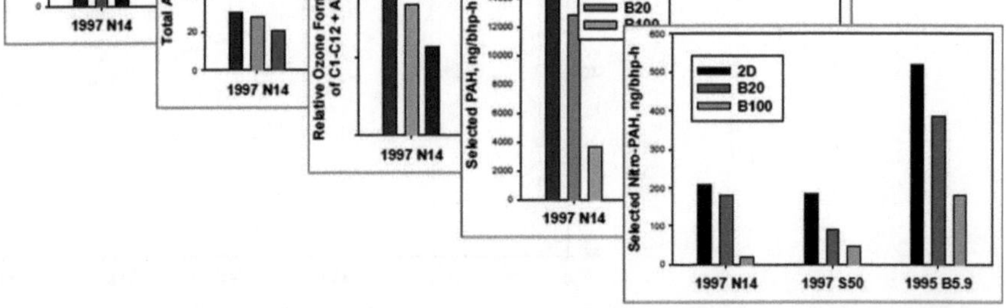

Fig. 17 Toxic emissions from biodiesel [11]

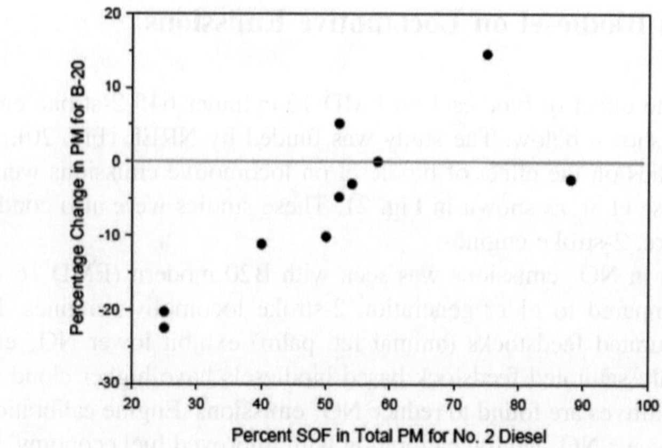

Fig. 18 Effect of SOF on PM from B20 fuelled engine [9]

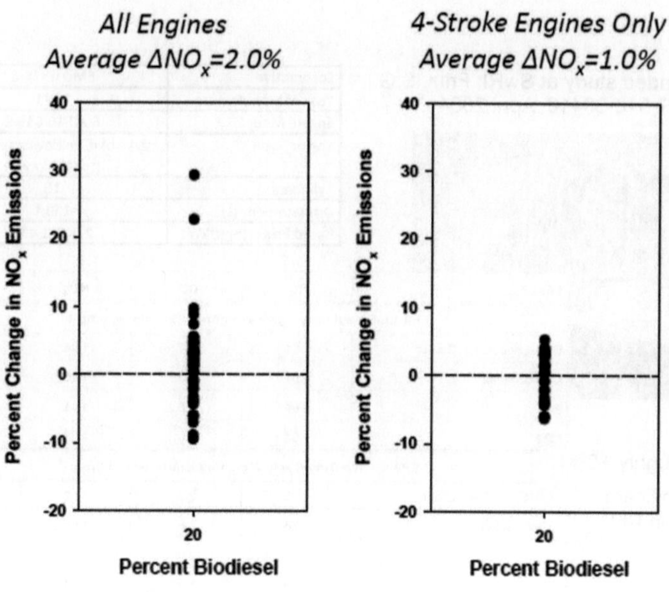

Fig. 19 Effect of 2-stroke engines on NO_x emissions [9]

Figure 19 illustrates the effect of engine cycle on NO_x emissions with B20 blend. Range of NO_x change was higher in 2-stroke engines compared to 4-stroke engines.

4 Effect of Biodiesel on Locomotive Emissions

Fritz studied the effect of biodiesel on EMD 16-cylinder 645 2-stroke engine and the results are shown below. The study was funded by NREL (Fig. 20).

Similar studies on the effect of biodiesel on locomotive emissions were carried out by Marchese et al. as shown in Fig. 21. These studies were also conducted on EMD large bore, 2-stroke engines.

A reduction in NO_x emissions was seen with B20 modern (EMD 16–710 GB) engines as compared to older generation 2-stroke locomotive engines. Biodiesel from more saturated feedstocks (animal fat, palm) exhibit lower NO_x emissions. However, highly saturated feedstock based biodiesels have higher cloud and pour points. Fuel additives are found to reduce NO_x emissions. Engine calibration can be altered to produce a NO_x neutral calibration with improved fuel economy. Selective Catalytic Reduction (SCR) systems have been successfully used on truck engines to reduce tailpipe NO_x emissions.

NREL funded study at SwRI: Fritz, S.G., NREL/SR-510-33416, April 2004

Locomotive	EMD GP38-2
Year of Manufacture	1973
Engine Model	EMD 16-645-E
Engine Type	2-Stroke, uniflow scavenged (roots blower)
Cylinders	16
Displacement (L)	169.1
Rated Power(bhp/kW)	2,000/1,491

Fuel	HC	CO	NO_x	PM
EPA Line-Haul Duty-Cycle Weighted Emissions, g/hp-hr [a]				
EPA Loco. Cert. Diesel	0.64	5.4	12.4	0.46
CARB diesel	0.64	4.3	12.3	0.46
B20	0.64	4.5	13.1	0.50
C20	0.64	4.0	12.8	0.48
EPA Switch Duty-Cycle Weighted Emissions, g/hp-hr [a]				
A Loco. Cert. Diesel	0.82	2.2	12.8	0.38
RB diesel	0.76	1.8	12.5	0.34
B20	0.78	2.0	13.5	0.37
C20	0.73	1.8	13.1	0.37

[a] Average of three runs on each fuel.

•NO_x roughly +5%
•No significant change in PM

Fig. 20 Effect of B20 on emissions from a 2-stroke large bore locomotive engine [12]

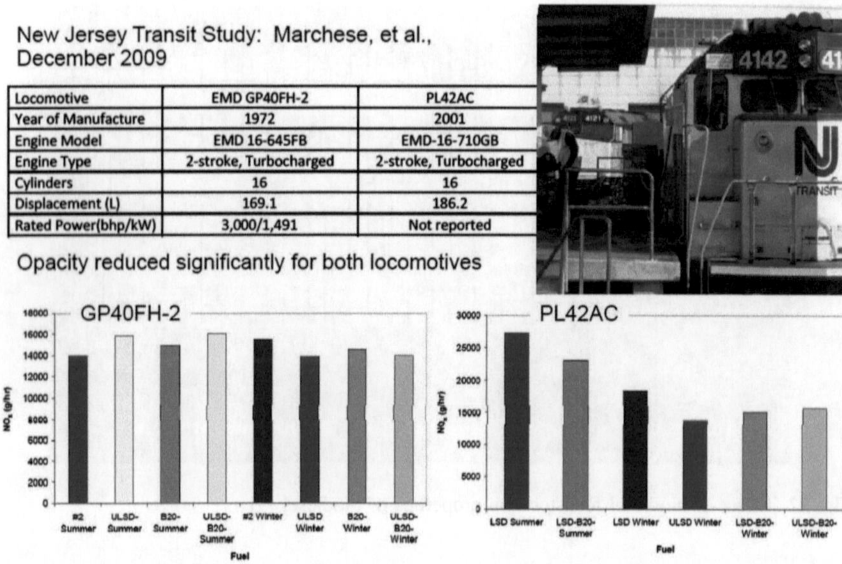

Fig. 21 Effect of B20 on emissions from a 2-stroke large bore locomotive engine [13]

5 Impact of Biodiesel on Lubrication of Diesel Engines

Gary M. Parsons of Chevron Oronite Company, USA [9] deliberated on the subject of lubricity effects of biodiesel on diesel engine. Heavy-duty OEM biodiesel issues and concerns can be divided into three areas namely: fuel system, emissions systems and lubricant performance. Injector deposits, fuel filter plugging, injection pump durability, materials compatibility, fuel instability, low temperature handling and reduction of detergency and anti-foam properties of fuel additive packages are some of the concerns related to the fuel systems. Impact on after-treatment devices and sensors, impact on NO_x emissions and lower BTU content and fuel economy are the issues related to performance and emissions from biodiesel fuelled engines. On the lubricants side, fuel dilution, corrosion, viscosity increase, oxidation, piston deposits and crankcase deposits are the subjects needing study and attention. Biodiesel feedstock properties impact performance of select biodiesels are shown in Figs. 22, 23, 24 and 25.

Figure 22 shows the effect of constituents of biodiesel (saturated vs. unsaturated fats) and blend percentage on cetane number, cold flow properties and oxidation stability of biodiesel blends for different feedstocks based biodiesel. It is seen that sunflower biodiesel has low saturated fats and high polyunsaturated fats, whereas coconut oil based biodiesel has high saturated fats and very low unsaturated fats. Other triglyceride based biodiesels have constituents inbetween these two extremes. It is shown in Fig. 22 that cetane value increases with higher amount of saturated

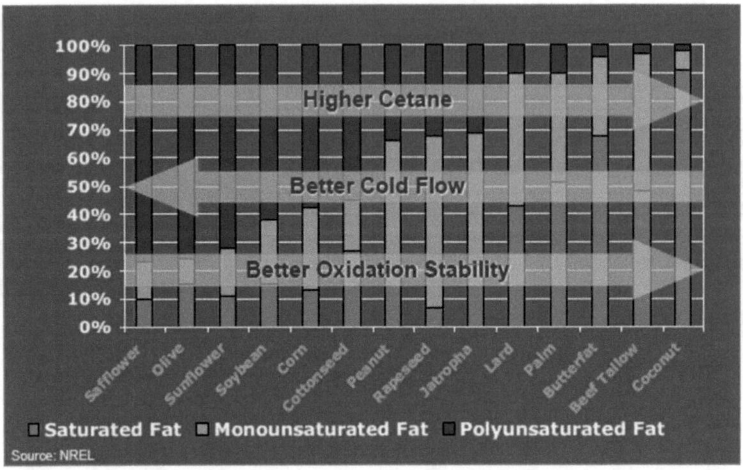

Fig. 22 Effect of biodiesel feedstock on properties of biodiesel [9]

Fig. 23 Effect of biodiesel on conventional diesel engine oils [9]

fats based esters in biodiesel leading to better oxidation stability. However the cold flow properties of biodiesel worsen as the amount of saturated fats in triglycerides increase.

Another effect seen with use of biodiesel is generation of engine deposits and lead corrosion. Biodiesel is known to react with metals like lead, brass bronze etc. and this may be a reason with lead deposits from the components containing lead. In some cases, additional deposits are seen on the pistons with use of biodiesel. These effects are illustrated in Fig. 23.

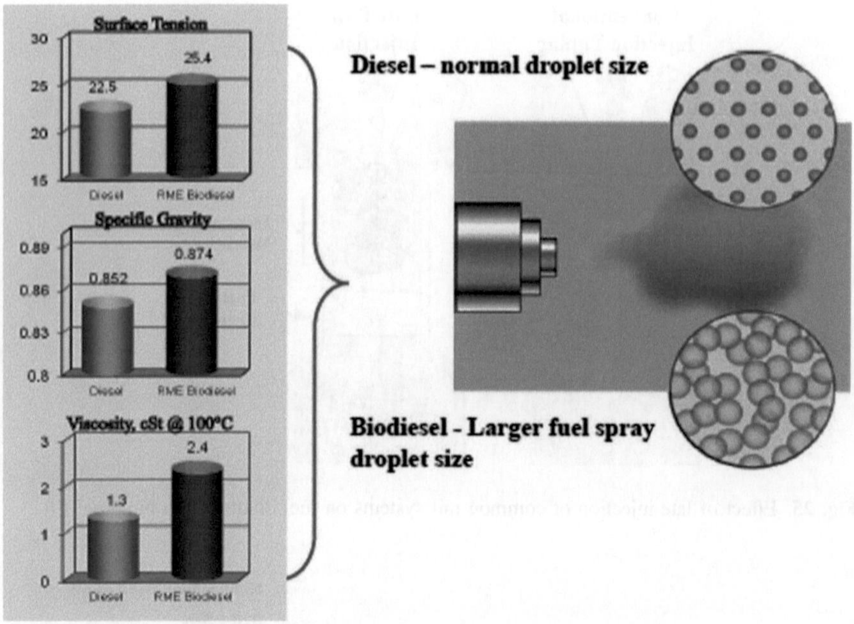

Fig. 24 Properties of Biodiesel lead to higher fuel dilution levels than mineral diesel [9]

Figure 24 shows the effect of biodiesel on biodiesel properties. Surface tension, specific gravity and viscosity of fuel increases with increased use of biodiesel. Also, the sauter mean diameter (SMD) and mean diameter of droplets increase with biodiesel content in the blend. This is expected to affect the spray characteristics as well as droplet evaporation time in the combustion chamber thus changing the combustion characteristics of the engine.

Figure 25 shows the effect of late post injection with common rail direct injection system on fuel dilution. Due to higher viscosity, density and larger droplet size, there is a tendency to have increased wall wetting, which adds to engine lube oil dilution. Some OEMs have reported 15–20% fuel dilution (FAME) at oil drain, hence OEMs often reduce drain interval when biodiesel is used (Fig. 30). Piston cleanliness and deposits are also reported to increase with soybased biodiesel (Fig. 26).

Figure 27 shows piston demerits (cleanliness and deposits) with use of biodiesel, diesel and different mixtures of biodiesel and Fischer-Tropsch diesel. B20 RME exhibited higher piston merit compared to ULSD but other blends deteriorated piston merits on use of biodiesel.

FAME biodiesel will likely remain a part of global diesel pool in future. Legal mandates, energy independence and support to local agriculture will become important drivers for introduction of biodiesel as compared to fuel economics. Production of FAME can be done with low capital investment and low complexity plants. In addition, new FAME technology is being developed. Use of biodiesel can

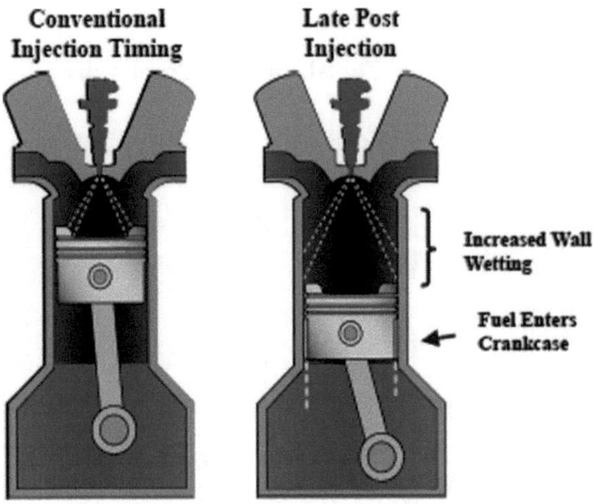

Fig. 25 Effect of late injection of common rail systems on fuel dilution with biodiesel [9]

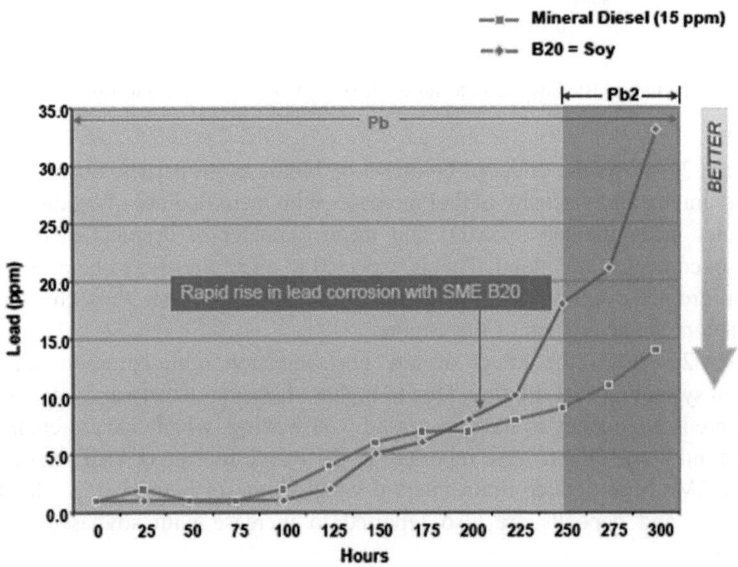

Fig. 26 Oil oxidation and lead corrosion identified with biodiesel blend [9]

lead to lubrication issues, high and persistent fuel dilution, oil oxidation, bearing corrosion and piston deposits. Specific methods have been developed for biodiesel dilution detection in used engine oil. Specially formulated lubricants can enable safe biodiesel use. OEMs are accepting greater use of biodiesel in engines due to introduction of BQ9000, ASTMD6751-091, and ASTM D7467-09a specifications.

6 Studies Carried Out at Engine Development Directorate of RDSO

In line with the guidelines issued by Government of India, RDSO has carried out engine characterization studies with non-edible feedstock based biodiesels on both large-bore 4-stroke and 2-stroke locomotive engines. Use of locally available feedstocks for production of biodiesel has a positive influence on agriculture, energy security and strengthening of local and national economy. At the same time, it is necessary to address the fuel vs. food issue, since a large population of India lives below the poverty line. Therefore selection of suitable feedstock to fulfill the above national objectives became another motivating factor for undertaking this research on locomotive diesel engines. Locally available non-edible raw materials such as cotton seed oil, pongamia seed oil, waste fish oil and mahua seed oils were chosen for conversion into biodiesels.

Non-edible character of these oils was the first consideration in their favor. Second consideration was their abundance in India as compared to other suitable non-edible feedstocks. Availability of these oils in India is presented in Table 2. Yearly production of biodiesel from these locally available non-edible oils can be in the range of 500–600 million liters, which is sufficient for IR to introduce at least 20% blend of biodiesel (B20) on its entire fleet of diesel locomotives.

All experimental studies were carried out at different locomotive engine notches for both engines. Figure 33 shows the torque and power produced by the two engines at these engine notches. Measurement of fuel injection, performance, emissions and combustion parameters was done at these notches. Comparison of engine characteristics was done for various test fuels on the two locomotive engines, i.e. 4-cycle and 2-cycle engines (Fig. 28).

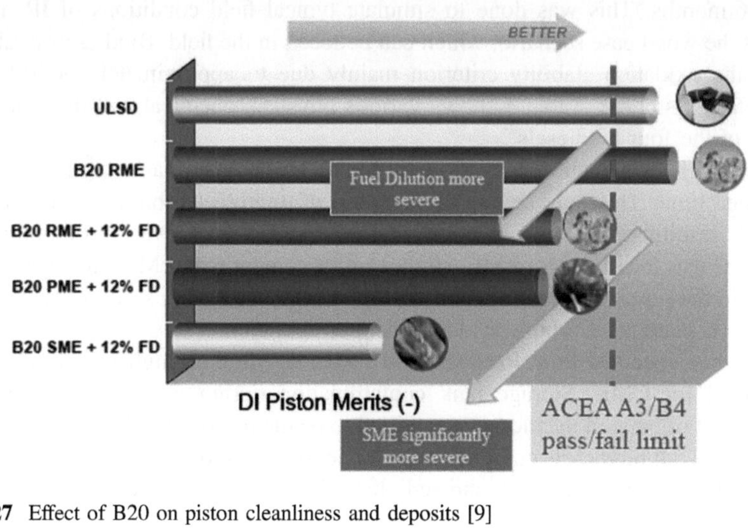

Fig. 27 Effect of B20 on piston cleanliness and deposits [9]

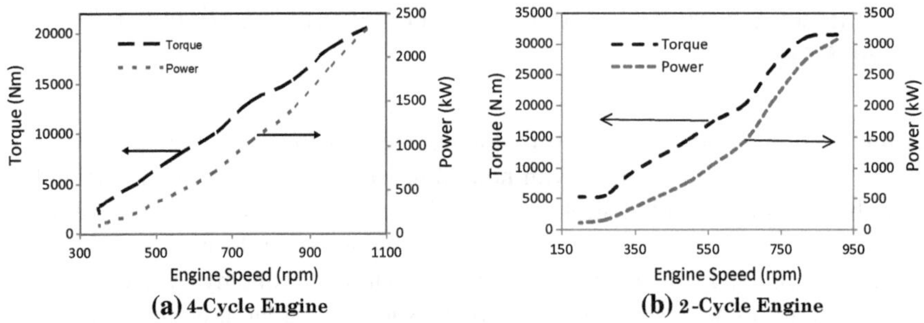

Fig. 28 Engine load lines for the two different locomotive engines [4]

6.1 Biodiesel Characterization

Four different biodiesels used in the present study (Cottonseed methyl ester-CME, Pongamia methyl ester-PME, Mahua methyl ester-MME and Fish oil methyl ester-FME) are shown in Fig. 34. Besides the differences in physical, chemical and thermal properties, differences in color and texture of these biodiesels can be clearly seen in Fig. 34. This is due to different feedstocks used for production of these biodiesels. Smell of these biodiesels is similar to the raw feedstock oils. Biodiesel fuelled engine exhaust also has the mild odor of original feedstocks.

Biodiesels samples were characterized in accordance to the Indian biodiesel standard [IS: 15607(2005)]. Additional tests were carried out as per the European Standards for biodiesel [EN 14214]. All biodiesel samples complied with majority of biodiesel specifications prescribed by the standards although there were some minor deviations. These biodiesels were tested after storage in ordinary tanks without any nitrogen cover and without addition of any antioxidant for a period over 6 months. This was done to simulate typical field conditions of IR and to assess the worst case scenario, which can be faced in the field. Biodiesels could not meet the oxidation stability criterion mainly due to approximately 5000 h long unprotected storage. Table 3 shows various physical, chemical and thermal properties of the four biodiesels.

For FME, kinematic viscosity was found to be marginally higher than the specified limit. This may be due to oxidation of unprotected biodiesel over a long period, resulting in polymerization of fuel molecules. Lower flash point than specified was measured for FME. Higher sulfur content with FME can be attributed partly to the possible contamination during storage and higher sulfur content of the feedstock material. Carbon residue and total contamination were observed to be higher than specified limit for FME and CME. In addition, higher acid value was measured for CME. Storage tank contamination during prolonged storage and reaction of organic molecules are suspected to be the reasons for this. For rest of the parameters, all biodiesels complied with the legislative requirements. The results of TLC-FID analysis (Table 4) showed that biodiesels contained 99% or greater

Table 3 Physical, chemical and thermal properties of biodiesels

Property	Unit	Test method	Limit (IS: 15607 (2005))	FME	CME	MME	PME
Density @ 15 °C	kg/m^3	ISO 3675, ISO 12185	860–900	889	887	889.3	880
Kinematic viscosity @ 40 °C	cSt	ISO 3104	2.5–6.0	**6.05**	5.44	5.5	5.178
Flash point (PMCC)[a]	(°C) min	IS 1448 P:21	120	**76.6**	144	149	157
Sulphur	(mg/kg) max	ASTM D-5453	50	**96**	42	13	26
Carbon residue (Ramsbottom)	(% w/w)	ASTM D-4530/ISO 10370	0.05	**0.36**	**0.18**	0.048	0.05
Sulphated ash	(% w/w)	ISO 6245	0.02	0.009	0.01	0.032	0.018
Water/water and sediment	ppm	ASTM D-2709, ISO 3733, ISO 6296	500	Nil	<500	<500	480
Total contamination	(mg/kg) max	EN 12662	24	**36.4**	**Membrane choked during test**	21	20
Cu corrosion	(3 h @ 50 °C) max	ISO 2160	1	1	1	1	1
Cetane no	Min	IS 5156	51	54.1	56	60.7	54
Acid value	(mg KOH/gm) max	IS-1448, P:1/Sec 1	0.5	0.47	**0.71**	0.26	0.061
Methanol	(% w/w) max	EN 14110	0.2	0.003	0.095	0.03	0.025
Free glycerol	(% w/w) max	ASTM D-6584	0.02	0.00063	0.0034	0.001	Nil
Total glycerol	(% w/w) max	ASTM D-6584	0.25	0.0037	0.032	0.018	0.0329
Phosphorous	(mg/kg) max	ASTM D-4951	10	<1	<1	<10	1
Oxidation stability, @110 °C	(hr) min	EN 14112	6	**1.54**	**2.21**	**5.45**	**1.27**
Sodium and potassium	(mg/kg) max	EN 14108 and EN 14109	To report	<1 and <1	2 and <1	<1 and <1	11 and 0
Calcium and magnesium	(mg/kg) max	EN 14108 and EN 14109	To report	<1 and <1	<1 and <1	<1 and <1	1 and 1
Iodine value		NMR method	To report	101.7	98.8	–[b]	–[b]

(continued)

Table 3 (continued)

Property	Unit	Test method	Limit (IS: 15607 (2005))	FME	CME	MME	PME
Monoglyceride content	(% w/w) max	ASTM D 6584	0.8 (as per EN 14105)	0.0286	0.0031	0.201	0.031
Diglyceride content	(% w/w) max	ASTM D 6584	0.2 (as per EN 14105)	Nil	Nil	Nil	Nil
Triglyceride content	(% w/w) max	ASTM D 6584	0.2 (as per EN 14105)	Nil	Nil	Nil	Nil

[a]Pensky Martin Closed Cup
[b]Iodine value (IV) could not be measured for these two biodiesels due to technical problem in the equipment. An estimation of IV can be made from GC analysis and degree of unsaturation of the constituent esters of biodiesels

Table 4 Composition of biodiesel samples

Saturated/Unsaturated	Fatty acid	Name	Structure	CME	FME	PME	MME
				% (w/w)			
Saturated	C14:0	Methyl Myristate	$C_{15}H_{30}O_2$	2.5	6.5	–	–
Saturated	C16:0	Methyl Palmitate	$C_{17}H_{34}O_2$	29.1	34.6	11.1	25.5
Mono-unsaturated	C16:1	Methyl Palmitoleate	$C_{17}H_{32}O_2$	2.2	6.6	–	–
Saturated	C18:0	Methyl Stearate	$C_{19}H_{38}O_2$	5.6	4.2	4.9	13.6
Mono-unsaturated	C18:1	Methyl Oleate	$C_{19}H_{36}O_2$	19	17.8	59.7	47
Poly-unsaturated	C18:2	Methyl Linoleate	$C_{19}H_{34}O_2$	41.6	30.8	24.3	13.9

methyl esters. Deviations observed from the specifications however did not affect the combustion of biodiesel in the engine. In addition, the engine has three tier fuel filtration arrangement therefore these minor deviations were not expected to influence the present study of locomotive engine performance, emissions and combustion significantly.

6.1.1 Gas Chromatography

Gas chromatography of biodiesels was done to characterize the constituent esters as per ASTM D1983. This test method establishes standard conditions for separation and identification of methyl esters by gas-liquid chromatography and is applicable to animal fat and vegetable oil fatty acids and oils having 8–24 carbon atoms. The use of polyester liquid phase facilitates the separation of both, saturated and

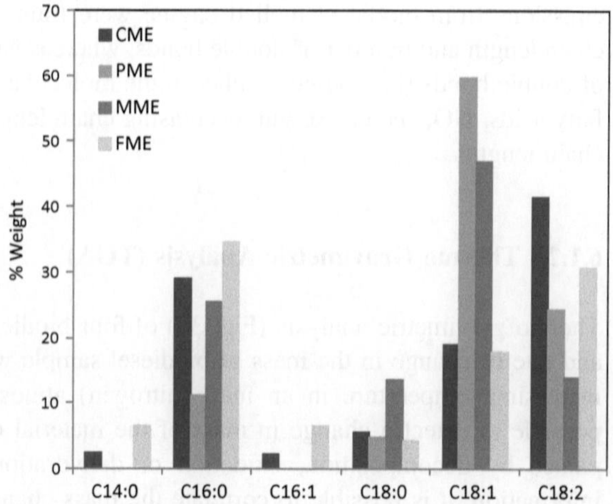

Fig. 29 Gas liquid chromatography results of biodiesel samples

Table 5 Normal boiling points of various methyl esters [16]

Acid chain	Myristic	Palmitic	Palmitoleic	Stearic	Oleic	Linoleic
T_b (°C)	296	338	394.2	352	349	366

unsaturated fatty acid methyl esters on the chromatogram obtained. Results of the GC analysis are presented in Fig. 29 and Table 4. Six fatty acid methyl esters (Fig. 29) ranging from 12 carbon to 18 carbon chain length with presence of saturated, mono-unsaturated and poly-unsaturated molecules were the constituents of biodiesels used in this research.

CME consists of primarily poly-unsaturated linoleic acid (Table 5) and palmitic acid methyl ester with relatively lesser amount of mono-unsaturated oleic acid methyl ester. CME also contains small quantities of stearic acid, myristic acid and palmitoleic acid methyl esters. FME consists of mainly palmitic acid, linoleic acid and oleic acid methyl esters. Lesser percentage of palmitoleic acid, myristic acid and stearic acid methyl esters are seen in FME. Higher presence of oleic acid methyl esters followed by linoleic acids and smaller quantity of palmitic and stearic acid methyl esters are observed in PME. MME has predominance of oleic acid and palmitic acid methyl ester with smaller quantities of stearic acid and linoleic acid methyl esters.

Thus biodiesels are mainly composed of three methyl esters, methyl palmitate (11.1–34.6%), methyl oleate (17.8–59.7%) and methyl linoleate (13.9–41.6%) (Table 5). Methyl oleate is considered to be a suitable major component of biodiesel for reducing NO_x emissions. Methyl palmitoleate has considerable advantages compared to methyl oleate, for low-temperature flow properties. It has been reported that SOF concentration and ignition delay increase with a reduction in fraction of methyl oleate ester and increase in the fraction of methyl linoleate ester in biodiesel. This is due to low ignitability of methyl linoleate ester; NO_x and PM emissions from biodiesel fuelled engine were found to be functions of fatty acid chain length and number of double bonds, whereas NO_x increased with the number of double bonds (i.e. iodine number in the molecular structure). For fully saturated fatty acids, NO_x increased with decreasing chain length from 18, 16 and 12 carbon chain lengths.

6.1.2 Thermo Gravimetric Analysis (TGA)

Thermo gravimetric analysis (Fig. 30) of four biodiesels was carried out. Amount and rate of change in the mass of biodiesel sample was measured as a function of increasing temperature in an inert (nitrogen) atmosphere. With TGA, it is also possible to detect a change in mass of the material due to phase changes accompanied by decomposition, oxidation or dehydration of the material. With this information, it is possible to correlate the mass changes to the chemical structure, processing and end-use performance.

Differences in the distillation curves of four biodiesels need to be studied in conjunction with the normal boiling points of their constituent esters. Normal boiling points of pure esters, which are the constituents of the biodiesels under investigation, are shown in Table 5.

PME has the lowest weight loss (% w/w) and CME, MME and FME have higher weight loss over the entire temperature range (200–500 °C). PME has the largest percentage of methyl oleate (60% w/w) and has its normal boiling in the upper range (349 °C), second only to the methyl linoleate. CME, MME and FME have the largest percentages of methyl palmitate, which is a low normal boiling point ester. Thus the distillation curve of biodiesels is a function of the normal boiling points of the ester constituents.

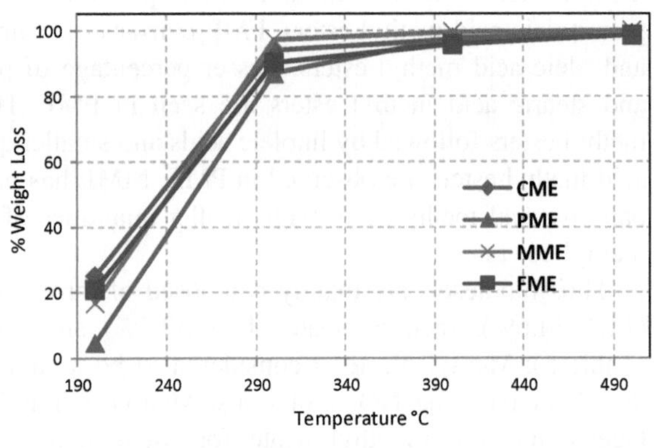

Fig. 30 TGA of biodiesel samples

Table 6 Heating values of the test fuels

Fuel	MME	CME	FME	PME	Diesel
LHV (MJ/kg)	41.3	41.4	42.7	41.3	46.2

PME with a higher mass percentage of methyl oleate shows lower evaporation up to 300 °C. On a volume basis. 75% (v/v) distillation of biodiesel takes place by 365 °C as compared to 289 °C for mineral diesel. Droplet evaporation inside the combustion chamber is affected by the volatility of fuel. Diesel with a lower evaporation temperature is therefore expected to show a faster vaporization inside the combustion chamber compared to biodiesel. This will affect the combustion of fuels, engine performance and emissions.

6.1.3 Calorific Value

Table 6 shows the heating values of fuels characterized in present study. The heating value of a fuel indicates the heat of reaction at constant pressure or constant volume at 298 K for complete combustion of unit mass of fuel. It is the energy that is released during combustion per unit mass of fuel. For hydrocarbons, the phase of H_2O formed during combustion affects the heating value. Energy released in fuel with the condensation of water vapors to liquid in the exhaust gas gives higher heating value (HHV). LHV is a more direct measure of the energy contained in the fuel and is used when water remains in vapor form. LHV of the biodiesels is ~10% lesser than mineral diesel due to presence of oxygen atoms in the fuel molecules. This has a direct influence on the amount of fuel required to produce a certain horsepower output from the engine.

Viscosity, density and cetane number of biodiesel samples are higher than mineral diesel, therefore they lead to different fuel injection characteristics, formation of spray inside the combustion chamber and auto-ignition vis-à-vis mineral diesel. Viscosity, cetane number and calorific value of biodiesels increase with carbon chain length and decrease with unsaturation. Different composition of biodiesels therefore determines combustion, performance and emissions from an engine. The range of these variations depends on the relative amount of various esters present in the test fuel.

6.2 Engine Performance Characterization

Figure 31 shows the power output, fuel injected per stroke per cylinder, BSFC and BTE of the engine with different test fuels. The control system of the test cell is able to maintain the required power output at each engine notch for all fuels (Fig. 31a–c). Fuel injected per stroke per cylinder shows a variation for different fuels and increases as the % blend of biodiesel increases (Fig. 31d–e) in the test blend. At 8th

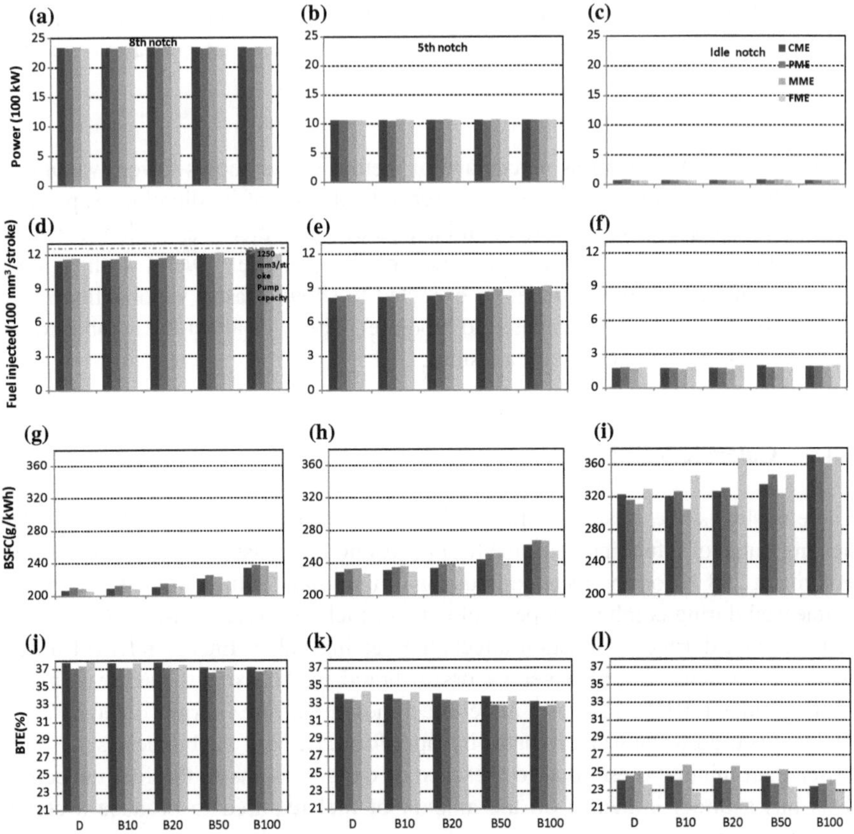

Fig. 31 Comparison of power output, fuel injected per stroke per cylinder, BSFC and BTE of the locomotive engine fuelled with different biodiesel blends

engine notch, MME and PME have the highest fuel injected per stroke per cylinder, almost reaching the limit (1250 mm^3/stroke) of the fuel injection pump used on this test engine. Thus B100 fuelled engine rated at higher than 2313 kW may show loss of power output, if fuelled with MME or PME. It will be essential to redesign fuel injection pump to allow higher volumetric flow rate of test fuels. Higher volumetric capacity of the fuel injection pump is an important consideration in design of biodiesel fuelled engines, especially if higher blends of biodiesel or B100 is planned to be used.

BSFC increases for higher blends or pure biodiesels as fuel (Fig. 31g–i). There is a strong variability in the BSFC for different biodiesels. At idle engine notch, B10 and B20 FME exhibit an unusually high BSFC and lower BTE as compared to other fuels. This needs further investigation. Examination of BTE obtained for different fuels reveals (Fig. 31j–l) that at 8th notch, highest BTE values were obtained. This is expected because the fuel injection timings are optimized for the

8th engine notch and fuel injection pump does not have any mechanism to vary the fuel injection timing with engine load and speed.

6.3 Engine Emission Characterization

Mass emissions from the locomotive engine with different fuels are shown in Fig. 32. Regulated emissions of NO_X, PM, THC and CO as well as CO_2 emissions were measured in ppm and converted to brake specific emissions by making use of procedure outlined in ISO-8178.

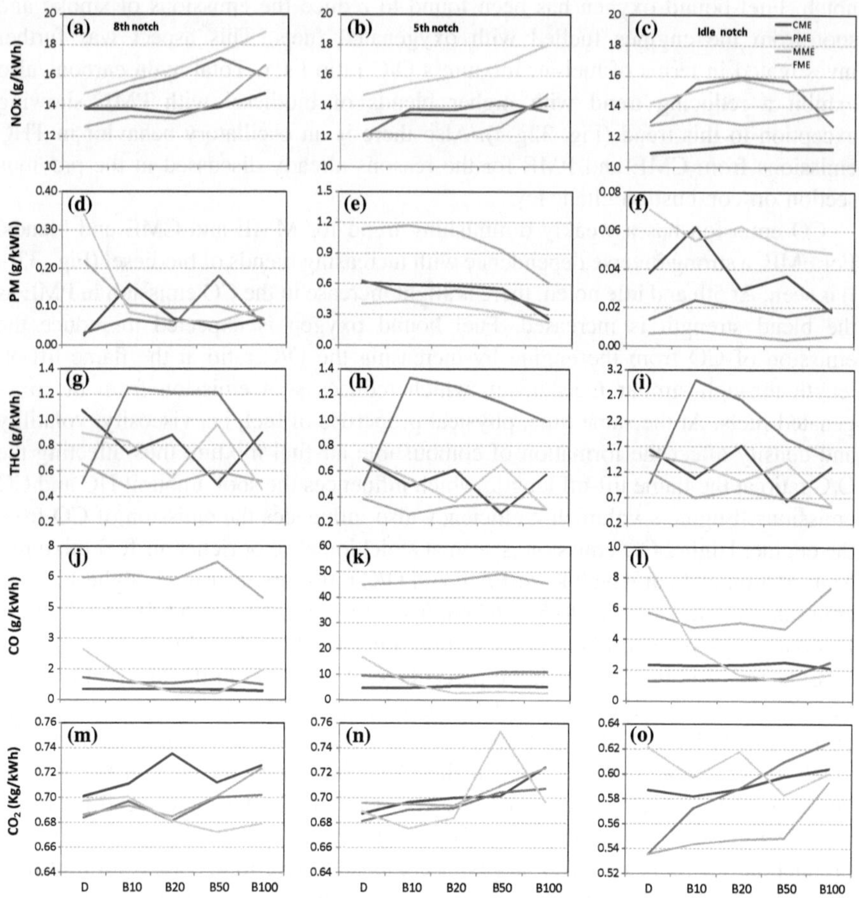

Fig. 32 Mass emissions from the test engine for various test fuels at three representative engine notches

At 8th notch, $BSNO_x$ emissions tend to increase with increase in biodiesel percentage in the test fuel (Fig. 32a). For PME, however there is a drop in $BSNO_x$ emissions between B50 and PME. For CME blends, B20 fuelled engine has the lowest $BSNO_x$ emissions. Similarly at 5th engine notch, a rising trend of $BSNO_x$ with increasing biodiesel blends is seen, PME being an exception. $BSNO_x$ for PME is lesser than B50. At idle engine notch, biodiesel fuelled engine is seen to emit higher $BSNO_x$ emissions than diesel for CME, FME and MME. For CME and FME, B20 emits highest $BSNO_x$ emissions whereas B20 MME emits lowest. For PME, blend (B10, B20 and B50) fuelled engine have considerably higher $BSNO_x$ emissions than diesel and biodiesel. Higher $BSNO_x$ from biodiesel fuelled engines vis-à-vis diesel fuelled engines has also been reported by other researchers.

Emission of particulate matter from the engine decrease, as the amount of biodiesel in the fuel increases (Fig. 32d–f). This trend is observed at three representative engine notches, with PME and blends showing an exception at 8th engine notch. Fuel bound oxygen has been found to reduce the emissions of smoke and soot from the engines fuelled with oxygenated fuels. This aspect was further investigated in terms of fuel-air mixture's O/C ratio later. Total hydrocarbons also exhibit a reducing trend with higher blends of biodiesel with PME showing exception to this trend (Fig. 32g–i). Also there is an oscillatory behavior in THC emissions from CME and FME for the reasons already discussed in the previous section on combustion efficiency.

CO emission has a weakly diminishing trend for MME and CME and blends. For FME, a strong inverse dependence with increasing blends of biodiesel (Fig. 32j–l) is seen. At 5th and idle notch, there is slight increase in the CO emission in PME as the blend strength is increased. Fuel bound oxygen is expected to reduce the emission of CO from the engine by increasing the O/C ratio at the flame lift-off length through similar mechanism, which reduces soot emission from the oxygenated fuels. At the same time, physical properties of fuel, i.e. viscosity, volatility and density affect the formation of combustible air-fuel mixture thus affecting the O/C ratio at the flame lift-off length, which influences the soot, unburnt HC and CO emissions. Engine's volumetric efficiency also influences the emission of CO from the engine. Higher CO emission is seen at stoichiometric or richer air fuel mixtures. Thus at 5th notch, at which lowest A/F and O/C ratios are measured, highest BSCO emissions are measured up to 50 g/kWh. In comparison, 8th notch CO emissions are less than 7 g/kWh and idle notch CO emissions less than 9 g/kWh.

CO_2 emission increase with the percentage of biodiesel in the test fuel and this trend is seen at the three representative notches (Fig. 32m–o). However, there are exceptions. FME fuelled engine has reducing CO_2 emissions at 8th and idle engine notch as the amount of biodiesel in the blend increases. Increase in the $BSCO_2$ emission of biodiesels is due to early carboxylation of ester moiety and relatively more complete combustion of air–fuel mixture. Szybist et al. found that CO_2 production occurs in the early stages of combustion due to decomposition of fuel, whereas this phenomenon could not be seen in a diesel surrogate fuel 'n-heptane'. Thus, CO_2 production in biodiesel fueled engines follows additional pathways other than oxidation of carbon.

7 Fuel Equipment Manufacturer's Joint Statement on Biodiesel [14]

Fuel equipment manufacturers like Delphi, Bosch, Denso, Continentla and Stanadyne have taken the following position with regard to use of biodiesel in engines equipped with their fuel equipment. Use of B5 according to ASTM D 975-09 is accepted by the FIE manufacturers although the absence of stability requirement is seen as a large risk that should be reduced as soon as possible. Blends containing in excess of 5% (v/v) FAAE (ASTM D 6751-09) require positive validation of specific issues associated with higher concentrations of low stability FAME.

FAME Stability: Lower oxidation stability of FAME is of particular concern. Aged or poor quality FAME contains organic acids like formic acid and polymerisation products. The acids attack many components and the polymers, which can lead to plugged filters, sediments and sticking moving parts, drastically reducing the service life of the FIE.

FAME Impurities: As FAME is being produced from an increasing number of new feedstocks, uncertainties are associated with additional impurities that might not become evident until vehicle operation.

FAME compatibility: Material compatibility issues may occur on older vehicles, designed before the use of FAME was considered. As FAME concentration increases, compatibility issues become more likely, especially with filters, hoses and seals being the most likely components affected.

Other Biofuels: FIE manufacturers support the use of bioparaffins obtained by hydro-treatment or co-processing of plant oil. Due to their paraffinic nature and high fuel and transport system compatibility, bioparaffins are also well suited for blending with biogenic portion above 7%.

Unestrified plant oil: The FIE manufacturers do not agree with the use of unesterified plant oil, even where such fuel meets existing national standards such as DIN V 51605.

8 Well to Wheels Energy Usage of Different Fuels [15]

Argonne National Laboratory (ANL) has carried out well-to-wheels energy usage of different fuels including biodiesel. This is shown in Fig. 33.

ANL has followed different methods in carrying out this analysis and used their in-house 'GREET' program to do the calculations. The pump to wheel energy usage does not vary for different fuels, however the wheel to pump energy usage is different for different fuels. Wheel to pump energy usage is calculated based on the displacement, energy allocation, market allocation and hybrid methods. The WTP energy usage for biodiesels is similar to the mineral diesel and in some cases lesser also. Similarly ANL has analysed Well to Wheel fossil energy usage of different

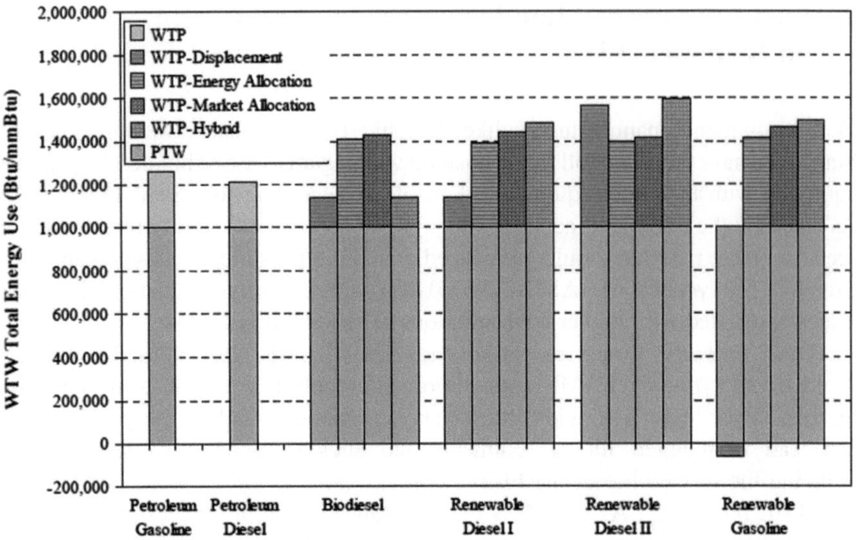

Fig. 33 WTW total energy usage of different fuels [15]

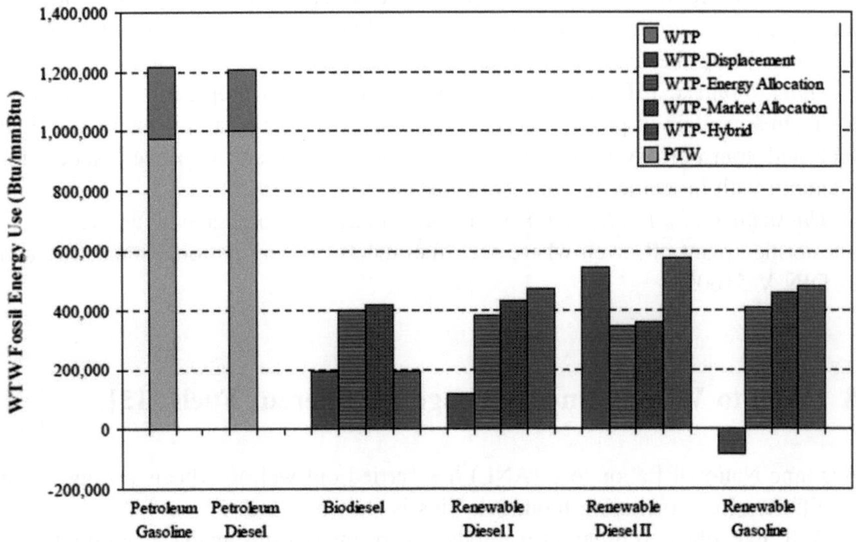

Fig. 34 WTW fossil energy usage of different fuels [15]

fuels including biodiesel (Fig. 34). Biodiesel showed the lowest fossil energy usage amongst all the fuels.

9 Costing

With deregulation of diesel fuel price, costing of biodiesel can follow a more realistic patch. IR has started purchasing biodiesel from Indian Biodiesel producers at prices that are either closely matching with that of diesel or are vary competitive. A comparable price to diesel fuel will provide sustainability to biodiesel albeit subsidies and support by the GoI may still be needed. Energy crops concept may positively impact the cost of raw material as has been proven in countries like Malaysia and Indonesia.

10 Roadmap for Indian Railways

Procurement and project management is the first priority for IR. IR is required to relook at its procurement procedures to have preferred partners and reduce delays in its procurement cycle. IR is required to set up blending mechanism for B5 at its fuelling points. There is an immediate need for creating supply logistics infrastructure, and quality control and monitoring infrastructure. RDSO should carry out durability studies for introducing higher blends of biodiesel (upto B20). The Industry on its side should consider the logistics in collection of already available oil-seeds and energy crop promotion. We should progressively reduce dependence on imported raw material and increase the share of indigenous bio-oils. Industry is also required to consider adequate in-process auditing and compliance with BQ 9000 standard requirements.

11 Summary

In this chapter, we have looked at various dimensions of using biodiesel as a transport fuel for Indian Railways. It is feasible to use lower blends of biodiesel without any major modifications in the locomotive engine however the storage, handling and usage issues need to be addressed in its full ernest. For using higher blends of biodiesel, detailed engine studies to ascertain engine and systems level effects are required, durability testing being one of them. However, the expected issues arising out of these studies are addressable and may require engine modifications. Overall, biodiesel has proved to be a feasible alternate locomotive fuel and can be implemented rather quickly.

References

1. Biofuels policy (2009) Government of India
2. Abbott D (2010) Keeping the energy debate clean: how do we supply the world's energy needs?. Proc IEEE 98(1)
3. Agarwal AK (2007) Biofuels (alcohols and biodiesel) applications as fuels for internal combustion engines. Prog Energy Combust Sci 33:233–271
4. Gautam A (2013) Experimental investigations of comparative performance, emission, and combustion characteristics of biodiesel-fueled four cycle and two cycle locomotive diesel engines. PhD Thesis. IIT Kanpur
5. Huber GW (2008) Breaking the chemical and engineering barriers to ligno-cellulosic biofuels: next generation hydrocarbon biorefineries. National Science Foundation, Chemical Bioengineering, Environmental and Transport Systems Division, Washington D.C
6. Huber GW, Iborra S, Corma A (2006) Synthesis of transportation fuels from biomass: chemistry, catalysts, and engineering. Chem Rev 106:4044–4098
7. www.virent.com
8. BQ 9000 Biodiesel standard (USA)
9. Proceedings of the SAE TC 7 sub-commitee on biodiesel (2010)
10. Engine manufacturers statement on use of biodiesels (2009)
11. Sharp C, Howell S, Jobe J (2000) The effect of biodiesel fuels on transient emissions from modern diesel engines, part ii unregulated emissions and chemical characterization. SAE Technical Paper 2000-01-1968. doi:10.4271/2000-01-1968
12. Fritz SG (2004) NREL/SR-510-33416
13. Marchese Anthony J, Bhatia Krishan K, Hesketh Robert P, McKenna David (2009) Evaluation of emissions and performance of NJ TRANSIT diesel locomotives with B20 biodiesel blends. Rowan University
14. Fuel equipments manufacturers common statement on use of biofuels (2009)
15. Huo H, Wang M, Bloyd C, Putsche V (2008) Life-cycle assessment of energy and greenhouse gas effects of soybean-derived biodiesel and renewable fuels. Report no. ANL/ESD/08-2, Argonne National Laboratory
16. http://webbook.nist.gov/

Fuel Properties and Emission Characteristics of Dimethyl Ether in a Diesel Engine

Hyun Gu Roh and Chang Sik Lee

Abstract The purpose of this article is to provide the physicochemical properties and characteristics of exhaust emissions from a DME-fueled diesel engine. Fundamental properties of DME are described and compared with the properties of the conventional diesel fuel. In addition, the applicability of DME fuel is presented in terms of its potentials as an alternative fuel for compression ignition engines. The exhaust emission characteristics of a DME-fueled engine are described using experimental results of different DME and diesel blends. The most important properties of DME are the presence of oxygen in the fuel molecule, lower ignition temperature, and high cetane number. The kinematic viscosity of DME is significantly lower than that of diesel. However, low lower hating value (LHV), low viscosity, and low lubricity are some of the major disadvantages of DME. The bulk modulus of DME is about 1/3 compared to diesel fuel. Spray tip penetration of DME is found to be shorter than that of diesel. Also, the Sauter mean diameter (SMD) distribution of the DME spray is found to be much smaller than that of diesel. Emission characteristics of different DME and diesel blends are investigated, which shows low soot, CO, and other hydrocarbon emissions.

1 Introduction

In the transportation and industrial fields, petroleum is the basic energy source for the power generation of vehicles and industrial plants. In particular, energy supply and security issues have become important energy problems such as fuel depletion and energy saving plan over the world. According to the projection of primary energy consumption until 2035 [1], the industrial sector is the dominant consumer of primary energy, and the second major component is the energy consumed by other

H.G. Roh
Mechanical and Automotive Engineering, Induk University, Seoul 01878, Republic of Korea

C.S. Lee (✉)
School of Mechanical Engineering, Hanyang University, Seoul 04763, Republic of Korea
e-mail: cslee@hanyang.ac.kr

A.K. Agarwal et al. (eds.), *Locomotives and Rail Road Transportation*,
DOI 10.1007/978-981-10-3788-7_6

sectors including residential service and agricultural sectors. The transportation sector is projected the third field to contribute to the growth in primary energy consumption. The application of diesel engines has significantly grown because of its high thermal efficiency and lower fuel consumption compared to gasoline fueled engines. The regulation of greenhouse gas is getting more severe. In recent years, the reduction of CO_2 emission and lower fuel consumption for transportation vehicles have been investigated the primary focus towards the research and development in the area of engine combustion. According to the report on CO_2 emissions from fuel combustion in 2015 [2], two sectors produced nearly two-thirds of global CO_2 emissions in 2013: electricity and heat generation accounted for 42% and transport accounted for 23%. Therefore, in order to study various substitute energy sources, many researchers are investigating on alternative fuels for the compression-ignition (CI) engines. From this point of view, one of the most expected future fuels is dimethyl ether (DME), which is an organic compound with the chemical formula CH_3OCH_3.

DME is a colorless, non-toxic, slightly narcotic, and highly flammable gas under ambient condition, but it can be handled as a liquid fuel at slightly pressurized condition [3, 4]. Among several alternative fuels, DME is a clean fuel, which produces a very less amount of pollutants compared to other conventional fuels. It has a high cetane number, unique autoignition characteristics, near-similar thermal efficiency and output performance of diesel engines [5–7]. It is a renewable fuel, which can be produced from any feedstock, such as biomass, coal, petroleum, solid wastes, etc. It has similar physical properties of liquefied petroleum gas (LPG) [4], therefore it is considered a clean alternative fuel to diesel [8–10]. However, the properties and combustion characteristics of DME are necessary to investigate before it can be applied in actual vehicle engines. The benefits of DME as an alternative fuel for automotive diesel engines are higher oxygen content and no carbon–carbon bonding in its chemical structure [11–15], therefore it has suitable properties of a substitute fuel for conventional diesel in compression ignition engines. In the case of DME-fueled CI engine, although many experimental and theoretical studies on DME combustion and mixture formation were conducted, it still has difficult problems such as low calorific value, poor lubricating properties, and modification of fuel supply and high-pressure injection system and so on.

The purpose of this article is to provide the physicochemical properties and characteristics of exhaust emissions from DME-fueled engine. Fundamental properties of DME are presented and compared with the properties of the conventional diesel fuel. In addition, the applicability of DME is presented in term of its potentials as an alternative fuel for compression ignition engines. Finally, the exhaust emissions from DME-fueled engine are characterized using experimental results of pure diesel and DME-biodiesel blend.

2 Fuel Properties of DME

This section introduces the thermochemical properties and application characteristics of DME as an alternative fuel in diesel engines. In particular, the important properties and application characteristics of DME fuel are described: oxygen content, autoignition temperature, cetane number, viscosity, LHV, compressibility characteristics, and problems of CI engine application.

2.1 Thermochemical Properties

2.1.1 Oxygenated Fuel with Better Ignition Characteristics

The most important property of DME is the presence of oxygen atom in the fuel molecule. The oxygen content of 34.8% in DME plays an important role in promoting combustion and reducing the formation of soot and particulate matter (PM) emissions [4, 12, 16]. The oxygen content and the absence of carbon–carbon bonds in the fuel structure promote the reduction of PM and near-zero smoke emissions. DME contains the second highest oxygen content among various fuels as shown in Fig. 1.

This figure is plotted from the references data [17–21]. When DME is blended with diesel, the fuel blends DD1090, DD2080, DD3070 contain 3.44, 6.96, and 10.44% of oxygen, respectively. In the figure, DD1090 fuel is the blend of DME10% and diesel 90% by volume ratio, and DD2080 and DD3070 are DME 20, 30% and diesel 80, 70% blends, respectively. In a diesel engine, the cetane number is an important indicator for the self-ignition of fuel.

Table 1 summarizes the physical and chemical properties of DME and a conventional diesel fuel. As listed in Table 1, the cetane number of DME (55–60) is higher than that of diesel. The higher cetane number of DME results in the shorter premixed burning period and ignition delay, which promotes combustion characteristics and reduced exhaust emissions [14].

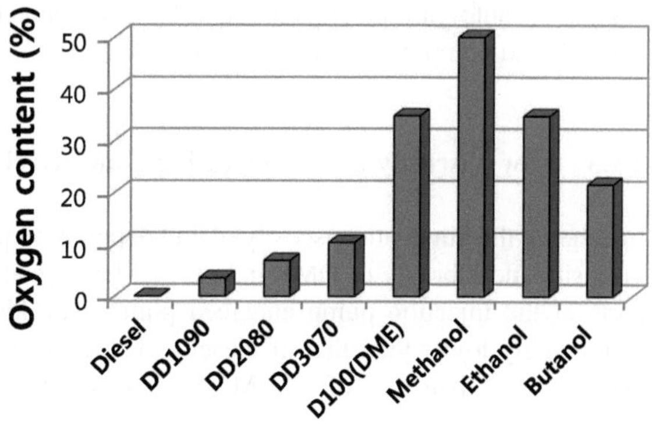

Fig. 1 Comparison of oxygen content for various fuels [17–21]

Table 1 Physical and chemical properties of DME and diesel fuels [1, 22]

Property	DME	Diesel
Chemical formula	CH_3OCH_3	C_8–C_{25}
Molecular weight	46.07	96 ~
Liquid density (at 20 °C) (kg/m^3)	660	800–840
Liquid viscosity (at 25 °C) (kg/ms)	0.12−0.15	2–4
Gas specific gravity (vs air)	1.59	–
Vapor pressure (at 20 °C) (bar)	5.1	<0.01
Boiling temperature (°C)	−25	≈150–380
Lower heating value (MJ/kg)	28.43	42.5
Enthalpy of vaporization at normal temperature and pressure	460(−20 °C)	250
Stoichiometric air fuel-ratio (kg/kg)	9.0	14.6
Cetane number	55–60	40–55
Modulus of elasticity (N/mm^2, 20 °C, 2 MPa)	553	1549

2.1.2 Autoignition Temperature

In the compression ignition engine, autoignition (or self-ignition) is a basic physical phenomenon in the engine cycle, and it is a dominant factor affecting the overall performance in a CI engine. In a diesel engine, the intake air is heated during the compression stroke. When fuel is injected into the high-temperature air, spontaneous ignition occurs after a short duration, which is usually referred as ignition delay. Ignition happens without a spark plug when the fuel temperature is raised above its autoignition temperature. The autoignition temperature of a fuel is the minimum temperature required to ignite the mixture in the normal atmosphere without an external source of ignition. The temperature at which an injected fuel ignites decreases as the oxygen concentration increases. The lower ignition temperature of DME is one of the major advantages compared to diesel fuel [22, 23]. Under atmospheric conditions, the autoignition temperature of DME is about 235 °C, which is slightly lower than that of conventional diesel (250 °C) [22]. However, the autoignition temperature of DME decreases with an increase in pressure because it depends on the ambient temperature. Under the typical operating pressure in a CI engine, the autoignition temperature of DME is found to be lower than that of diesel, ethanol, and compressed natural gas (CNG) [24].

2.1.3 Low Viscosity and Lubricating Characteristics

In general, the kinematic viscosity of fuel depends on pressure and temperature. Low viscosity and lubricity of DME induce fuel leakage and surface wear in the moving parts of the injection pump and feed pump. The kinematic viscosity of DME is significantly lower than that of diesel as listed in Table 1. This viscosity range of DME is less than that of ASTM standard specification for diesel fuel oil [25].

In order to improve the lubricating characteristics, DME is needed to be used along with some additives, such as Lubrizol, biodiesel, and other hydrocarbon fuels [8, 24, 26].

2.1.4 Low Heating Value and High Latent Heat of Vaporization

As listed in Table 1, the lower heating value (LHV) of DME is about 28.43 MJ/kg, which is about two-third of the LHV of diesel. Therefore, in order to compensate the low LHV of DME, the actual engine should be supplied with more fuel when it is operated with DME for the same power output. In actual CI engine operation, to improve the drawback of lower heat amount per unit mass of DME, it is necessary to blend higher LHV fuel with DME in a DME fueled engine. In order to obtain the same power output of the engine, at least 35% more DME is required to be introduced in the engine when compared to diesel.

On the other hand, the high latent heat of DME also offers an advantage as an alternative fuel. The latent heat of vaporization for DME is 460 kJ/kg, while that of diesel is about 250 kJ/kg at normal temperature and pressure [1]. In the case of DME-fueled engine, the higher latent heat reduces the NO_x emission because the injected DME spray absorbs heat from the ambient air in the cylinder during the mixing process with spray vaporization. Therefore, the DME-air mixture in the combustion chamber experiences a larger temperature drop compared to that of diesel-air mixture. When liquid DME injected into the charged high-temperature air in the combustion chamber, the low boiling point of DME results in the faster vaporization of fine droplets in the DME sprays compared to diesel sprays. Therefore, the fast vaporization due to the low boiling point of DME results in a decrease in temperature of premixed combustion.

2.1.5 Compressibility Characteristics

In a high-pressure injection system in a diesel engine, the bulk modulus of DME is an important property related to the generation of injection pressure and the propagation speed of pressure waves [27]. The modulus of elasticity of DME is about 1/3 of diesel fuel as listed in Table 1. Therefore, in case of DME engines, low modulus of elasticity of DME can bring about undesirable injection characteristics such as a decrease in injection quantity and delayed injection timing. Therefore, the DME injection system needs to be modified to meet the injection capacity. Comparing the temperature dependence of the bulk modulus, DME displays large temperature dependence of the bulk modulus than that of diesel fuel. Sato et al. [27] reported that when the temperature rises from 20 to 60 °C, the bulk modulus of DME drops to 37%, while that of diesel fuel drops to 19%. Therefore, the temperature dependence on the injection quantity of DME is much larger than that of conventional diesel.

2.1.6 Safety and Environmental Problems

DME is a volatile organic compound that is not corrosive to metals; however, some elastomers are not compatible with DME. The specific gravity of DME is very important on the side of safety. The specific gravity of DME is 1.59, and this value fall within the value of propane (1.52) and butane (2.01) [1]. As seen here, DME is heavier than the air, which may be a significant factor in actual applications from the safety perspective. Therefore, a good ventilation system in the work space is needed when DME is used. During the combustion process, DME burns with a visible blue flame, and DME smells like ether, and thus odorants addition are not required [28]. As an organic compound, DME is non-toxic, non-carcinogenic, non-teratogenic, non-mutagenic, and environmental benign [29].

In the study of lifetimes and global warming potentials (GWPs) of dimethyl ether and fluorinated, Good et al. [30] showed that the tropospheric lifetime of DME is 0.015 years and its GWP (time horizon) is 1.2 (for 20 years), 0.3 (for 100 years), and 0.1 (for 500 years). In addition, GWP is defined as the time-integrated warming effect caused by the instantaneous release of a unit mass (1 kg) of a greenhouse gas into the atmosphere relative to that of carbon dioxide. They showed that the GWP value of DME is benign for the atmosphere and should not contribute significantly to global warming. Comparing the lifetime of CO_2, DME, methane, and N_2O, the lifetimes are 1, 0.015, 12, and 114 years, respectively [30]. As shown in these data, DME has a relatively lower lifetime and GWP values than the other gases for all time horizons. Comparing the explosive limit of DME and diesel, DME has a relatively lower explosive limit (3.0–4.0% by volume) compared to that of diesel fuel (about 0.6%) [24, 31, 33]. On the other hand, lower flammability limits of DME-air changes from 3.5 to 3.75% when the initial pressure changes from 100 to 40 kPa. At the same time, the upper flammability limits are found to drop significantly from 19 to 12.5% [32]. DME has a visible flame when it burns in the air, and minimum ignition energy is about 0.29 mJ [1].

2.2 *Application Characteristics as an Automotive Fuel*

2.2.1 DME Supply Infrastructure

With the depleting petroleum reserve, DME and other alternative fuels are receiving a great deal of attention for the potential applications in the transportation sector. In general, an alternative fuel vehicle is a vehicle that is powered by a fuel other than the conventional fuels, such as, gasoline and diesel. Recently, a number of studies with DME as an alternate fuel in diesel engines have been increased significantly. DME burns more cleanly than the other alternative fuels and it does not form the soot emission, and it reduces carbon monoxide and unburned hydrocarbon emissions. As mentioned earlier, the physical properties of DME are similar to that of LPG (composed of butane and propane).

DME is a stable ether compound in the gaseous phase at the atmospheric conditions of 0.1 MPa and 298 K. However, DME changes phase from gaseous to liquid phase when DME is pressurized above 0.5 MPa at the standard atmospheric temperature [3, 4, 11]. Since DME has properties which are similar to LPG fuels, the existing infrastructure in the land and ocean based facilities for LPG can be used. Moreover, DME can be transported using the existing LPG tankers and the same receiving stations may be used with minor modifications to the packing seals, gaskets, and feeding pumps. Therefore, using the similar transportation facilities and infrastructure for storing DME, it may be possible to use the existing LPG filling stations to dispense DME. Therefore, the transition from LPG to DME could be less expensive since it does not require building a completely new DME infrastructure [30].

2.2.2 Application Characteristics as a Diesel Substitute

The saturation pressure of DME at 20 °C of ambient temperature is about 5.1 atmospheres as listed in Table 1. In order to feed the liquid DME to high-pressure injection pump, the low pressure primary pump in the storage tank delivers a constant flow rate of liquefied DME fuel to the high-pressure injection pump. If DME fuel exists in the lower pressure than its subcooled pressure at any instant state, the fuel in feed pump system and intake process of the pump contains the flashing vapor and partially gas phase bubble, therefore fuel vapor in the liquid fuel brings about the vapor locked or incomplete force feeding. To avoid the feed obstacle, the fuel in the storage tank is usually kept in the compressed liquid state.

Among the DME properties, lower volumetric energy content, lower viscosity and higher vapor pressure are very important design factors for the fuel passage, injection nozzle, the fuel feed and high-pressure injection pump. From the exhaust emissions perspective, the DME fuel burns cleanly in the CI engine with lower pollutant emissions compared to those operating with the conventional diesel [3, 4, 11, 12, 15]. DME is an oxygenated fuel and chemical compound with C-O-C linkage, while conventional diesel has not oxygen content. This advantage contributes the cleaner emissions without soot concentration compared to diesel fuel. Using the existing engine technology, DME produces the least amount of well-to-wheel greenhouse gas emissions compared to biodiesel, bionaptha, methanol, methane, and ethanol [30]. DME immediately evaporates upon injection into the combustion chamber, therefore high injection pressure is unnecessary as like diesel injection. Lower injection pressure of DME fuel means that lighter design and less expensive system for fuel injection can be manufactured.

Excellent results for DME fuel as a diesel substitute in the CI engine are summarized as follows: (a) Effective cetane number and oxygen content in the fuel molecule than conventional diesel fuel, (b) Less noise and smooth combustion than diesel engines, (c) Low injection pressure and low liquefied pressure, (d) Overall efficiency similar to that of conventional diesel, (e) Almost zero soot (particulate) emissions, (f) Better cold start and low temperature operation, (g) Excellent vaporization characteristics, (h) Easy on board characteristics, and (i) Fuel properties similar to LPG.

3 Combustion and Emissions Characteristics

This section describes the combustion and emission characteristics of DME fueled CI engine. The first part is composed of spray penetration, atomization characteristics and combustion performance, and the second part introduces the emissions characteristics of DME fueled diesel engine such as CO, HC, NO_x, and soot emissions.

3.1 Atomization and Combustion Characteristics of DME

In diesel engines, fuel spray and atomization process are very important for the improvement of combustion and emission characteristics. Figure 2 shows the spray tip penetration of DME and diesel, and the effect of ambient pressure on the spray tip penetration [34]. As shown in Fig. 2, the spray tip penetration of DME is found to be shorter than that of diesel.

Comparing the impingement timings on the spray chamber wall, diesel spray is faster than the DME spray because of rapid spray propagation due to a higher inertia of diesel. In Fig. 2b, the effect of ambient pressure on the penetration showed that the spray tip penetration decreases with an increase in the ambient air pressure. As illustrated in Fig. 2a, b, diesel spray penetration was found to be longer than that of DME. The main reason for this is caused by higher vaporization property and lower density of DME compared to that of diesel fuel. In the figure, the predicted values obtained using the empirical relation show similar trends as observed in the experimental measurements of spray penetration [35].

Figure 3 shows the comparison of the overall Sauter mean diameter (SMD) between DME and diesel. The overall SMD of DME under the same experimental condition is much smaller than that of diesel sprays. The droplet SMD of both DME and diesel is decreased with the injection time elapsed after the start of injection. In order to confirm the vaporization characteristics of DME, the predicted evaporation characteristics of DME and diesel are illustrated in Fig. 4. As shown in the vapor distribution, the vapor and mixture distribution of DME spray is broadly distributed in the chamber space. However, in the case of diesel injection, the resulting sprays do not sufficiently evaporate at the same crank angle because of relatively poor vaporization characteristics of diesel fuel compared to DME. Comparing both spray behaviors, DME spray provides homogeneous mixing of fuel and air compared to diesel spray.

Figure 5 shows the combustion pressure and crank angle diagram of DME-biodiesel blend DB8020 (DME80%, biodiesel 20% by volume) at an engine

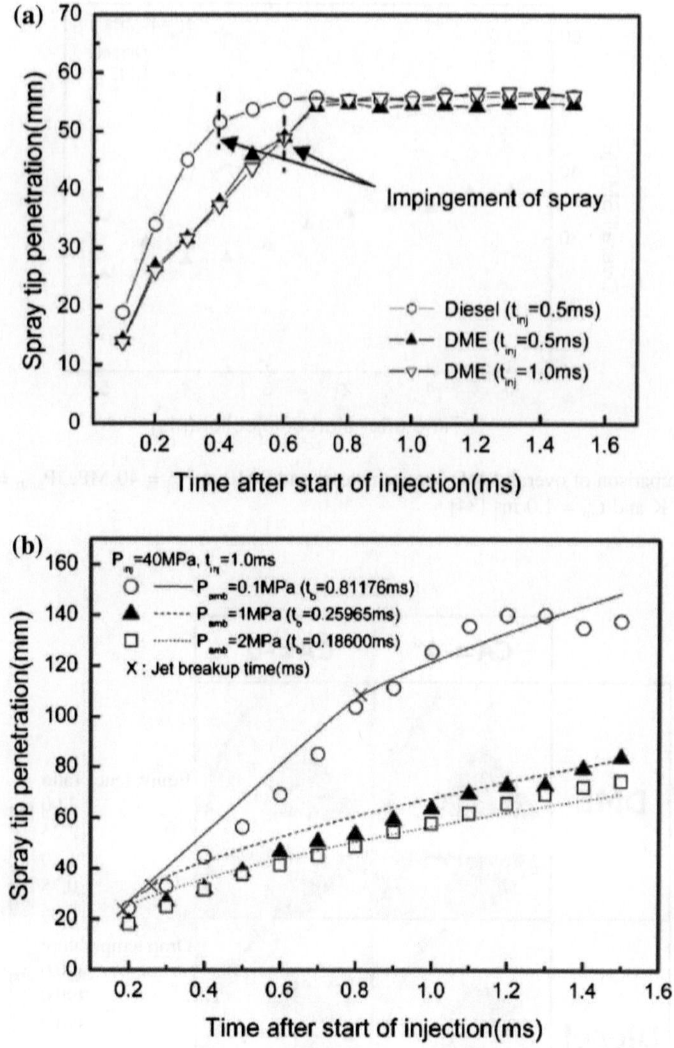

Fig. 2 Spray tip penetration of DME and diesel fuel at $P_{inj} = 40$ MPa and $P_{amb} = 0.1$ MPa (Injection duration 1.0 ms, ambient temperature = 293 K, *Symbol* Experiment, *Line* Empirical equation values) [34]. **a** Spray tip penetration and impingement timing. **b** Spray tip penetration according to ambient pressure

speed of 1500 rpm and 20 Nm of engine load. When DME is used in the engine, it is necessary to add a lubricating agent because DME has very low lubricating characteristics. In this work, biodiesel is an added improver for engine lubrication and it is a compensated fuel of heating value of DME. As mentioned earlier, the heating value of DME is about 28.43 MJ/kg, which is about two-third of diesel fuel, while the LHV of biodiesel is about 38 MJ/kg (89.4% of diesel) [36]. In the

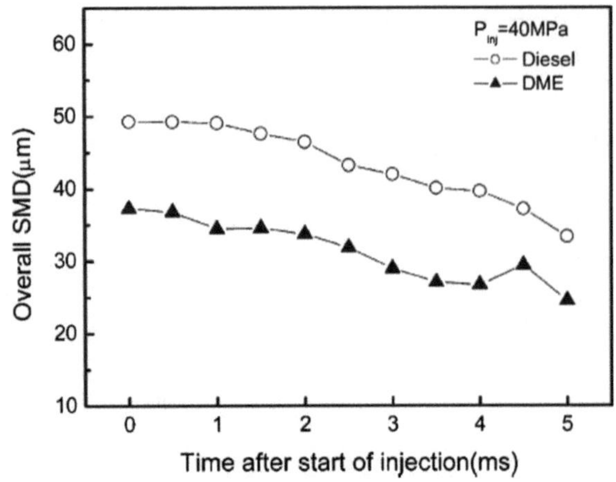

Fig. 3 Comparison of overall SMD between diesel and DME at P_{inj} = 40 MPa, P_{amb} = 0.1 MPa, T_{amb} = 293 K and t_{inj} = 1.0 ms [34]

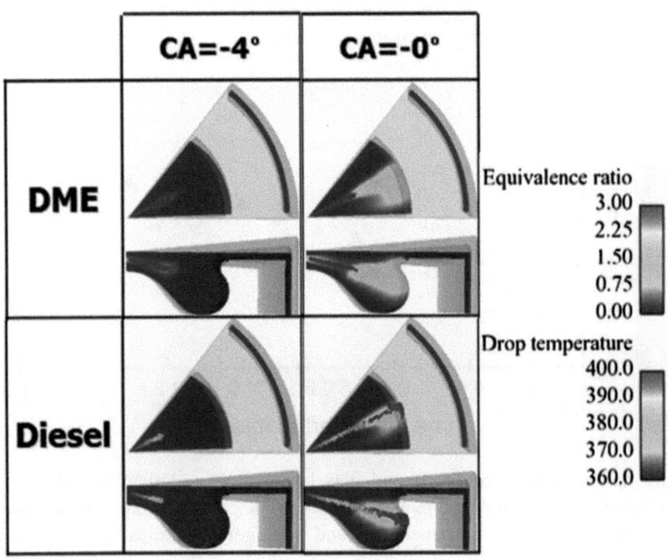

Fig. 4 Comparison of vapor distribution in the combustion chamber (Injection pressure = 50 MPa, Engine speed = 1500 rpm, Nozzle hole = 0.119 mm, Seven holes, Induced spray angle = 153°)

case of combustion pressure, DME-biodiesel blend shows a higher pressure than that of diesel. The main reason behind this trend is the better evaporation characteristics of DME, a high oxygen content in the fuel, a high cetane number and

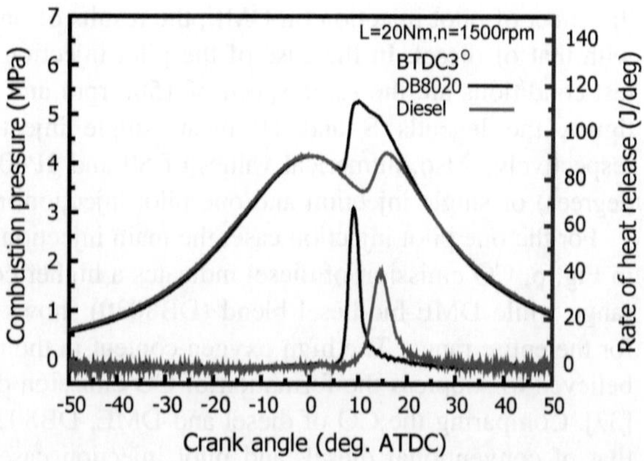

Fig. 5 Combustion characteristics of DME-biodiesel blend DB8020 and diesel fuel

improved heating value of blended fuel due to the mixing of biodiesel which influences the combustion improvement.

On the other hand, diesel shows the lower combustion pressure and the rate of heat release at the same engine load. This means that the injection near the top dead center (TDC) does not provide the sufficient time for fuel burning, which causes a lower combustion pressure and a retarded rate of heat release than those of the DME blend. DME showed the higher combustion characteristics than that of diesel because of its excellent vaporization characteristics and higher cetane value.

3.2 *Exhaust Emission Characteristics of DME*

Figure 6 illustrates the comparison of CO emission between DME-biodiesel blend (DB8020) and diesel for the single injection and one pilot injection cases. The emission results are obtained at various injection timings and 20 Nm of engine load for the single injection cases. In the case of Diesel and DB8020 fuels, the engine load is 20 Nm and fuel injection pressure is 50 MPa. Also, in order to investigate

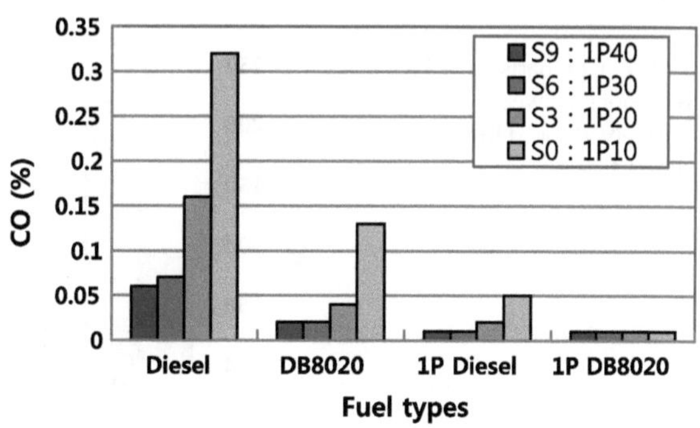

Fig. 6 Comparison of CO emission between DME-biodiesel DB8020 and diesel fuel

the effect of pilot injection on DME, the results of one pilot injection are compared with that of diesel. In the case of the pilot injection (1P Diesel and 1P DB8020), test conditions are the same speed of 1500 rpm and 60 Nm of engine load. In the figure, the legends S and 1P mean single injection and one pilot injection, respectively. Also, numerical values of S9 and 1P40 are injection timings (BTDC degrees) of single injection and one pilot injection, respectively.

For the one pilot injection case, the main injection timing is the TDC. As shown in Fig. 6, CO emission of diesel indicates a higher concentration for the entire test range while DME-biodiesel blend (DB8020) shows significantly lower emissions for the entire range. The high oxygen content in the molecular structure of DME is believed to suppress the formation of CO emission during the combustion process [37]. Comparing the CO of diesel and DME, DB8020 is remarkably reduced than that of conventional diesel, and pilot injection cases (1P Diesel, 1P DB8020) is more significantly reduced than those Diesel and DB8020 fuel.

In particular, the case 1P DB8020 shows very low level of CO emissions. Also, 1P Diesel reduces CO emission compared to the single injection case using diesel in the CI engine.

The comparison of HC emissions from the engine operated with DME-biodiesel blend and diesel is shown in Fig. 7. In the case of single injection, the HC emissions from DME-biodiesel blend (DB8020) significantly decreases compared to the HC emissions from diesel. The reduction in HC emission for DB8020 may be attributed to the high oxygen content and excellent evaporation characteristics of DME. On the other hand, results from the pilot injection and single injection show that pilot injection (1P DB8020) significantly reduces the HC emission compared to the single injection (DB8020). As stated earlier, DME is an oxygenated fuel and high cetane number than that of conventional diesel fuel. These properties help to promote the combustion characteristics, thus unburned hydrocarbon emissions are reduced. As mentioned earlier, the lower spray tip penetration, faster evaporation and ignition of DME, reduce fuel wall wetting on piston and combustion chamber surface, which is expected to result in lower HC emission.

Figure 8 shows the NO_x emissions using different DME-biodiesel blend when the engine is operated at 1500 rpm and at an injection pressure of 50 MPa. When DME is used in the CI engine, NO_x emission slightly increases at the same load condition compared to that of diesel combustion [1, 36]. In the advanced

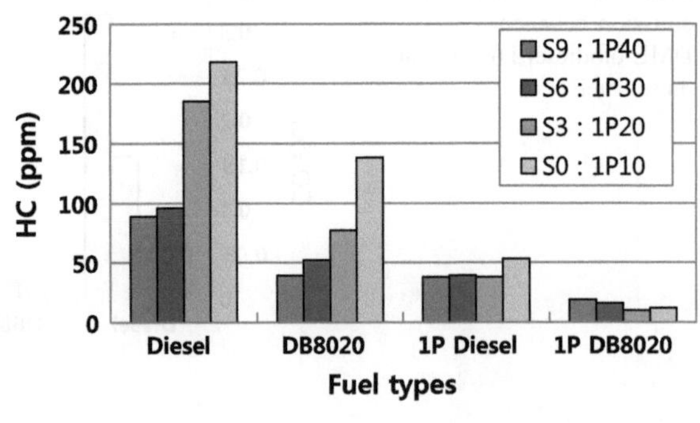

Fig. 7 Comparison of HC emissions between DME-biodiesel DB8020 and diesel fuel

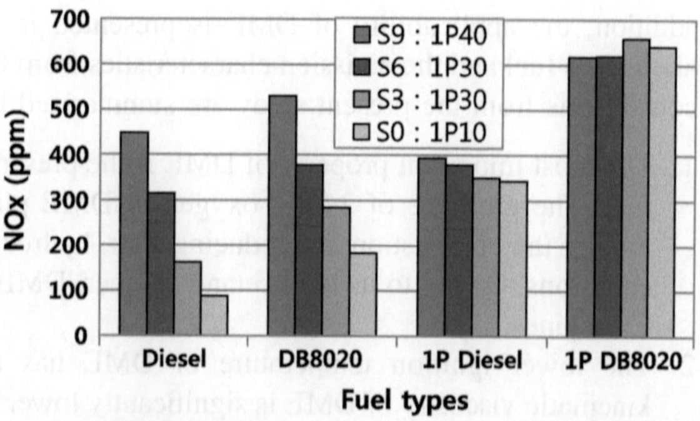

Fig. 8 Comparison of NO_x emissions between DME-biodiesel DB8020 and diesel fuel

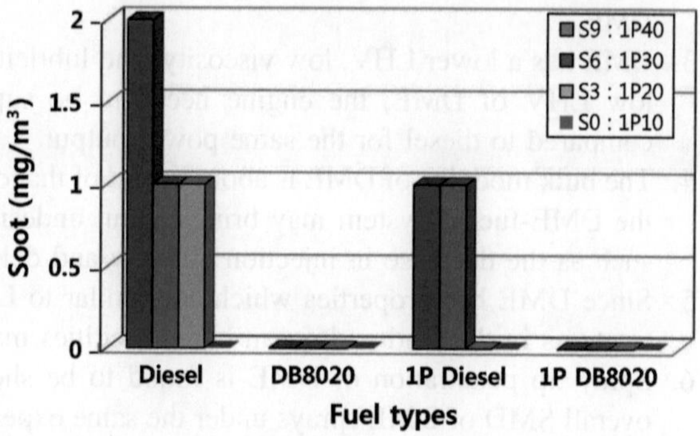

Fig. 9 Comparison of soot emissions between DME-biodiesel DB8020 and diesel fuel

injection timings, NO_x emissions are higher concentration for the single injection cases. It is attributed to faster ignition of the DME mixture due to advanced injection timing leading to an increased charge temperature. As a result, the elevated temperature produces more NO_x during combustion. In case of diesel, NO_x concentration is slightly lower than that of DB8020. In pilot injections (1P Diesel and 1P DB8020), the engine load is higher than the Diesel and DB8020, therefore, the combustion temperature increases.

The soot concentrations obtained using different test fuels are summarized in Fig. 9. The soot emissions using DB8020 and 1P DB8020 fuel show zero level of emission, while Diesel and 1P Diesel indicate high soot emission compared to the blend of DME-biodiesel (DB8020). The excellent vaporization, absence of carbon–carbon bond and high oxygen content in the fuel molecule of DME are resulted in lower CO and HC emissions.

4 Conclusions

The objectives of this chapter are to provide the physical and chemical properties, and emission characteristics of DME from a CI engine. Fundamental aspects of DME properties are described and compared with the properties of diesel. In addition, the applicability of DME is presented in terms of its potentials as an alternative fuel and the emission characteristics from CI engines. Some of the major conclusions from the present study are summarized below:

1. The most important property of DME is the presence of oxygen atom within the fuel. The presence of 34.8% oxygen in DME plays an important role in promoting the combustion and reducing soot, hydrocarbons, and carbon monoxide emissions. Owing to its high cetane number, DME is suitable for applications in CI engines.
2. The lower ignition temperature of DME has advantages over diesel. The kinematic viscosity of DME is significantly lower than that of diesel. In order to improve the lubricating characteristics, it is necessary to add some additives to DME.
3. DME has a lower LHV, low viscosity, and lubricity. In order to compensate the low LHV of DME, the engine needs to be supplied with more fuel when compared to diesel for the same power output.
4. The bulk modulus of DME is about a third of that compared to diesel. Therefore, the DME-fueled system may brings about undesirable injection characteristics such as the decrease in injection quantity and delayed injection timings.
5. Since DME has properties which are similar to LPG, the existing LPG infrastructures in the land and ocean based facilities may be utilized.
6. Spray tip penetration of DME is found to be shorter than that of diesel. The overall SMD of DME sprays under the same experimental condition is found to be much smaller than that of diesel sprays. In the case of combustion pressure, DME-biodiesel blend (DB8020) produces higher pressure than that of diesel.
7. CO emissions from diesel indicate a higher concentration for all test ranges while DME-biodiesel blend (DB8020) shows the significantly lower emissions. HC emission from the DME-fuelled engine is lower than the engine operated with Diesel.
8. When DME is introduced in a CI engine, NO_x emission increases compared to that of diesel combustion at the same engine load. The soot emissions using DB8020 and 1P DB8020 fuel show near-zero level, while Diesel and 1P Diesel show a significantly higher level of soot emission.

References

1. Park SH, Lee CS (2014) Applicability of dimethyl ether (DME) in a compression ignition engine as an alternative fuel. Energy Convers Manag 86:848–863
2. International Energy agency (2016) CO_2 emissions from fuel combustion. Highlights 2016:10–21
3. Semelsberger TA, Rodney LB, Howard LG (2006) Dimethyl ether (DME) as an alternative fuel. J Power Sources 156:497–511
4. Park SH, Lee CS (2013) Combustion performance and emission reduction characteristics of automotive DME engine system. Prog Energy Combust Sci 39:147–168

5. Denis G, Herwig O, Daniel S, Eddie S, Marc AW (2001) The performance of a heavy duty diesel engine with a production feasible DME injection system. SAE Tech Paper. 2001-01-3629
6. Kajitani S, Chen CL, Oguma M, Alam M, Rhee KT (1998), Direct injection diesel engine operated with propane-DME blend fuel. SAE Tech Paper. 982536
7. Kajitani S, Oguma M, Mori T (2000) DME fuel blends for low emission direct injection diesel engines. SAE Tech Paper. 2000-01-2004
8. Youn IM, Park SH, Roh HG, Lee CS (2011) Investigation on the fuel spray and emission reduction characteristics for dimethyl ether (DME) fuelled multi-cylinder diesel engine with common-rail injection system. Fuel Proc Technol 92:1280–1287
9. Crookes RJ, Bob-Manuel KDH (2007) RME or DME: a preferred alternative fuel option for future diesel engine operation. Energy Convers Manage 48:2971–2977
10. Kim HJ, Suh HK, Lee CS (2008) Numerical and experimental study on the comparison between diesel and dimethyl ether (DME) spray behaviors according to combustion chamber shape. Energy Fuel 22(4):2851–2560
11. Thomas G, Feng B, Veeraragavan A, Cleary MJ, Drinnan N (2014) Emissions from DME combustion in diesel engines and their implications on meeting future emission norms: a review. Fuel Process Technol 119: 286–304
12. Arcoumanis C, Bae C, Crookes R, Kinoshita E (2008) The potential of di-methyl ether (DME) as an alternative fuel for compression-ignition engines: a review. Fuel 87:1014–1030
13. Salvi BL, Subramanian KA, Panwar NL (2013) Alternative fuels for transportation vehicles: a technical review. Renew Sust Energ Rev 25:404–419
14. Teng H, McCandless JC, Schneyer JB (2003) Compression ignition delay (physical + chemical) of dimethyl ether—an alternative fuel for compression-ignition engines. SAE tech paper. 2003-01-0759
15. Roh HG, Lee D, Lee CS (2015) Impact of DME-biodiesel, diesel-biodiesel and diesel fuels on the combustion and emission reduction characteristics of a CI engine according to pilot and single injection strategies. J Energy Institut 88:376–385
16. Kim HJ, Park SH, Lee KS, Lee CS (2011) A study of spray strategies on improvement of engine performance and emissions reduction characteristics in a DME fueled diesel engine. Energy 36:1802–1813
17. Park SH, Yoon SH, Lee CS (2013) HC and CO emissions reduction by early injection strategy in a bioethanol blended diesel-fueled engine with a narrow angle injection system. Appl Energ 107:81–88
18. Wang Y, Zhou L, Wang H (2006) Diesel emission improvements by the use of oxygenated DME/diesel blend fuels. Atmos Environ 40:2313–2320
19. Nabi MN (2010) Theoretical investigation of engine thermal efficiency, adiabatic flame temperature. Appl Therm Eng 30:839–844
20. Suh HK, Lee CS (2016) A review on atomization and exhaust emissions of a biodiesel-fueled compression ignition engine. Renew Sust Rev 58:1601–1620
21. Yun H, Choi K, Lee CS (2016) Effects of biobutanol and biobutanol–diesel blends on combustion and emission characteristics in a passenger car diesel engine with pilot injection strategies. Energ Convers Manage 111:79–88
22. Yoon SH, Cha JP, Lee CS (2010) An investigation of the effects of spray angle and injection strategy on dimethyl ether (DME) combustion and exhaust emission characteristics in a common-rail diesel engine. Fuel Proc Technol 91:1364–1372
23. Huang ZH, Wang HW, Chen HY, Zhou LB, Jiang DM (1999) Study of combustion characteristics of a compression ignition engine fuelled with dimethyl ether. Proc Inst Mech Eng D—J Aut 213:647–652
24. Kapus P, Ofner H (1995) Development of fuel injection equipment and combustion system for Dl diesels operated on dimethyl ether. SAE Tech Paper. 950062
25. ASTM (2013) Standard specification for diesel fuel oils, ASTM 975-12
26. Christensen R, Sorenson SC, Jensen MG, Hansen KF (1997) Engine operation on dimethyl ether in a naturally aspirated. DI Diesel Engine, SAE Technical Paper. 971665

27. Sato Y, Nozaki S, Noda T (2004) The performance of a diesel engine for light duty truck using a jerk type in-line DME injection system. SAE Tech Paper. 2004-01-1862
28. Fleisch T, McCarthy C, Basu A, Udovich C, Charbonneau P, Slodowske W, Mikkelsen S-E, McCandless J (1995) A new clean diesel technology: demonstration of ULEV emissions on a navistar diesel engine fueled with dimethyl ether. SAE Tech Paper. 950061
29. Verbeek R, Van Der Weide J (1997) Global assessement of dimethyl ether: comparison with other fuels. SAE Tech Paper. 971607
30. Good DA, Francisco JS, Jain AK, Wuebbles DJ (1998) Lifetimes and global warming potentials for dimethyl ether and for fluorinated ethers: CH_3OCF_3 (E143a), CHF_2OCHF_2 (E134), CHF_2OCF_3 (E125). J Geophys Res 103(D21):28181–28188
31. Wakai K, Nishida K, Yoshizaki T, Hiroyasu H (1999) Ignition delays of DME and diesel fuel sprays injected by a D. I. Diesel injector. SAE Tech Paper. 1999-01-3600
32. Zhang B, Ng HD (2016) An experimental investigation of the explosion characteristics of dimethyl ether-air mixtures. Energy 107:1–8
33. Sorenson SC, Mikkelsen SE (1995) Performance and emissions of a 0.273 l direct injection diesel engine fuelled with neat dimethyl ether. SAE Tech Paper. 950064
34. Suh HK, Park SW, Lee CS (2006) Atomization characteristics of dimethyl ether fuel as an alternative fuel injected through a common rail injection system. Energy Fuels 20:1471–1481
35. Hiroyasu H, Arai M (1990) Structures of fuel sprays in diesel engines. SAE Tech Paper. 900475
36. Yoon SG, Lee CS (2011) Experimental investigation on the combustion and exhaust emission characteristics of biogas–biodiesel dual-fuel combustion in a CI engine. Fuel Process Technol 92:992–1000
37. Kim MY, Yoon SH, Ryu BW, Lee CS (2008) Combustion and emission characteristics of DME as an alternative fuel for compression ignition engines with a high pressure injection system. Fuel 87:2779–2786

Potential of DME and Methanol for Locomotive Traction in India: Opportunities, Technology Options and Challenges

Avinash Kumar Agarwal, Nikhil Sharma and Akhilendra Paratap Singh

Abstract Locomotives are critical for the economic development of India. Majority of locomotives of Indian Railways (IR) are diesel-electric locomotives. While these locomotives offer significant flexibility and ease of operation, they have significant challenges as well such as imported fuel consumption. As the fossil fuel resources are dwindling at an alarming rate, there is a need to explore alternative fuels such as DME and Methanol, whose potential for this applications largely remains unrecognized in India. Dependence on imported petroleum can be significantly reduced by these alternative fuels. In this chapter, technical feasibility, opportunities, challenges and potential of DME and Methanol as locomotive fuels in India is discussed. These fuels are low-emission fuels, which have been overlooked in both energy policy and industry discussions, despite many attributes, which make them attractive. DME is especially suitable in diesel locomotives due to its excellent fuel characteristics. Motivation to use DME as a candidate fuel in locomotives in India comes from its soot-free emission due to absence of C–C bonds. It can be also be blended with mineral diesel to overcome limitations of using pure DME, which include low viscosity and density. Similarly, Methanol can be blended with diesel for use in locomotives. Methanol has a high latent heat of vaporization; higher oxygen content, is sulfur free and has higher burning speeds than conventional liquid fuels. When burned at high temperature, it can reduce smoke and oxides of nitrogen (NO_x) emissions from compression ignition (CI) engines. However, there are limitations in the development of pure Methanol CI engines. Such limitations have been highlighted in this chapter and partial replacement of diesel by DME and Methanol are proposed as a solution for locomotives. Modifications to prevent corrosion and overcoming issues such as low lubricity in fuel injection systems is also discussed in detail.

Keywords Methanol · Di-methyl ether · Locomotives · Fuel injection equipment

---

A.K. Agarwal (✉) · N. Sharma · A.P. Singh
Engine Research Laboratory, Department of Mechanical Engineering,
Indian Institute of Technology Kanpur, Kanpur 208016, India
e-mail: akag@iitk.ac.in

A.K. Agarwal et al. (eds.), *Locomotives and Rail Road Transportation*,
DOI 10.1007/978-981-10-3788-7_7

1 Introduction

Indian Railways (IR) is the 'lifeline of India'. The first locomotive was operated between Bombay to Thane on 16th April 1853. Since then it has grown exponentially and has become an important asset and prime mover of the Indian economy. Today, it covers a total route length of more than 63,028 km. It comprises of 8000 locomotives, 50,000 coaching vehicles, 222,147 freight wagons, 6853 stations, 300 yards, 2300 good sheds, 700 repair shops, and has 1.54 million work force [1, 2]. IR runs approximately 11,000 trains every day, of which 7,000 are passenger trains. IR plays an important role in development of industries and helps the growth of the economy of the country. Table 1 shows the share of the transport sector in overall GDP in India.

1.1 Emissions from Locomotives

Locomotive engines contribute significantly to the air pollution in many cities and ports [4]. Locomotive emission standards and tests are well described in USEPA-420-R-98-101 (1998) and this is a benchmark for developing countries. Stringent emission norms imposed by the government have constantly increased the necessity of using clean fuels. Therefore, this is an appropriate time for India to look forward to new energy resources for locomotives, which can comply with upcoming stringent emission norms.

1.2 Conventional Fuels for Locomotive Traction

Fossil fuels play a critical role in developing the economy of a country since they are used to produce heat and power. Combustion of fossil fuels affects the environment through greenhouse gas (GHG) emissions. Annual expenditure on diesel consumed by locomotives of IR is roughly US $ 2.2 billion, which is ~4% of the

Table 1 Percentage share of transport sector in overall GDP [3]

	2008–09	2009–10	2010–11	2011–12	2012–13
Overall transport	6.6	6.6	6.5	6.6	6.7
Railways	1	1.0	1	1.0	0.9
Road transport	4.7	4.7	4.6	4.8	4.9
Water transport	0.2	0.2	0.2	0.2	0.2
Air transport	0.2	0.2	0.3	0.3	0.3
Services incidental to transport	0.4	0.4	0.4	0.4	0.4

Table 2 Fuel prices [3]

Item	2008–09	2009–10	2010–11	2011–12	2012–13	2013–14
Rs per kl of HSD	34,002	33,495	38,286	41,842	49,542	65,954
Growth over PY (%)	9.2	−1.5	14.3	9.3	18.4	33.1
Rs per kwh of EE	4.43	4.52	4.72	5.20	6.10	6.49
Growth over PY (%)	0.3	2.1	4.5	10.1	17.2	6.40

total diesel consumption of India, and is ~10% of the total revenue generated by IR [2]. Table 2 shows the fuel prices of high speed diesel (HSD) and electricity used to drive locomotives.

The growth over previous year (PY) is significant for both high speed diesel and electrical energy (EE). However price of HSD has increased every year at a much faster rate than electricity. Fuel bill has witnessed a steady increase on account of tariffs, consumption and increases activities. With ever growing concerns of the environmental pollution, energy security and unreliable future oil supplies, global community is seeking non-petroleum based alternative fuels along with more advanced energy technologies to increase the efficiency of energy usage in transport sector as well as in rail road sector.

1.3 Alternative Fuels for Locomotive Traction

Biomass, biodiesel and gasohol are promising alternative fuels for the future. Conventional fossil fuels have issues related to security of supply due to their limited availability. There are also issues related to above alternative biofuels usage such as limited natural resources, limited land available for cultivation of such crops and requirement of water for irrigation etc. To address such issues, there has to be identification of other liquid fuels, which are not derived from vegetable products or crops and can directly be used in existing engine using existing technologies. Availability, economics, acceptability, emission compliance, national security, technology, versatility etc. could be the criteria to judge the new alternative fuels.

Compared to other alternative fuels, DME and Methanol are emerging as leading alternative fuel candidates as they can sustain economic and environmental security for future generations. Their use will reduce dependence on petroleum as an energy source. Figure 1 shows the molecular structure of DME and Methanol.

Fig. 1 DME and Methanol molecular structures

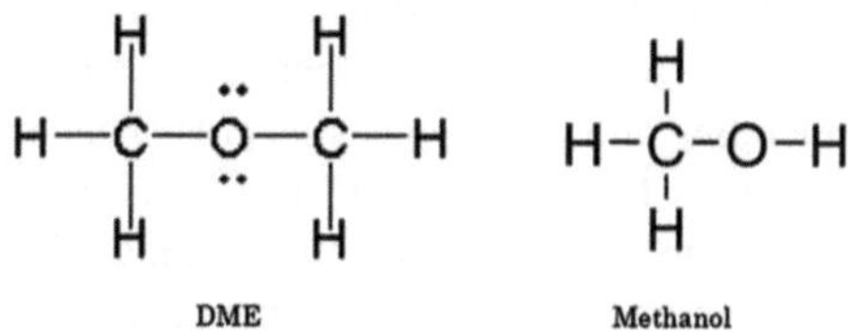

1.3.1 DME

Amongst synthetic energy resources, DME is emerging as the most appropriate near-term solution because of its superior engine and environmental performance [5–8]. It is being considered as a fuel for the 21st century. DME is a colorless, compressed gas, which is liquid at 5 bar pressure at room temperature. It is presently used as a refrigerant, solvent, propellant for aerosol sprays and in manufacturing certain plastics. DME is relatively non-toxic, although it is highly flammable [9]. It evaporates from water in an open system within few hours therefore alleviating the issues related to accidental leakage in water bodies. It degrades via a photochemical reaction with OH radicals and gets converted into carbon dioxide and water. DME can be synthesized from natural gas, coal, biomass and/or municipal solid waste (MSW), and is sulfur-free, near-zero aromatics content synthetic fuel, which is considered an excellent substitute for mineral diesel as well as liquefied petroleum gas (LPG).

1.3.2 Methanol

Methanol is also a promising alternative fuel, which can be blended with mineral diesel for use in locomotive engines. Unlike DME, Methanol is a highly volatile flammable, colorless, and toxic liquid fuel [10, 11]. Methanol is an oxygenated fuel and it helps in reducing emissions in a compression ignition (CI) engine. Low-carbon number alcohols such as Methanol have been encouraging as an alternative fuel due to their higher cetane rating than diesel. Characteristics of Methanol give it an extra edge over other alternate fuels as a suitable bio-alcohol in emerging biofuels market. Methanol is poisonous and affects central nervous system. Skin absorption and inhalation of Methanol leads to headache, blindness or serious visual impairment and even death [12]. Its symptoms are decreased level of consciousness, poor motion coordination, vomiting, abdominal pain, and a specific smell in the breath. Drinking small amount of Methanol may also lead to death. It is important precursor of CO, HCHO and O_3 [13]. Methanol produces lower PM and Sulphur dioxide emissions compared to diesel in a CI engine. Abundant availability and if domestically produced, it can reduce dependence on imported petroleum. This makes it a suitable fuel for locomotive application. However, Methanol increases the risk of corrosion, which must be dealt with by sufficient upgradation of fuel tanks and other relevant locomotive hardware.

Miscibility of Methanol with diesel is low, unstable and problematic because the two liquids separate very easily and an emulsifier or a co-solvent is required to stabilize the blend. However co-solvents or emulsifiers are expensive and the mixing process is cumbersome, which typically includes splashing, heating, blending and other processes [14]. This may be ascribed to great difference in their polarity. There are two main factors, which decide the quality of Methanol-diesel emulsion. These are the emulsification device and surfactants. Detailed information on blending diesel with Methanol was described in a research article published by

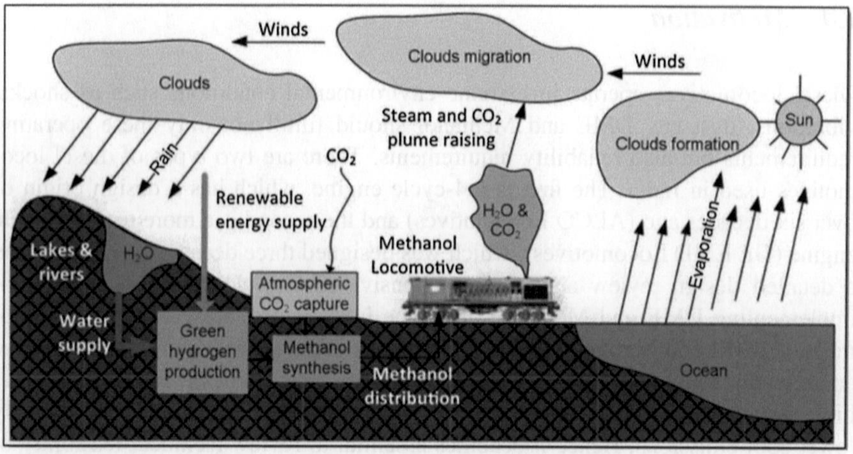

Fig. 2 Sustainable Methanol railway transport, integrated with water and carbon dioxide natural cycles [16]

Jiao et al. [15] wherein ways to achieve stabilization Methanol-diesel emulsion were described.

Figure 2 shows a sustainable model for Methanol based railway transportation system, integrated with water and carbon-dioxide natural cycles. This sustainable model shows entire cycle of Methanol production from natural resources for use in locomotives. Table 3 shows timeline of Methanol initiative with focus on railway transportation taken in USA, Brazil, South Africa and Canada. Methanol has properties similar to those of methane, when it is injected into the IC engine and can be used as a fuel in a dual-fuel engines.

Table 3 Methanol initiatives with focus on railway transportation undertaken globally [16–18]

Country	Year	Description
USA	1980	A direct injected, diesel-Methanol, two-stroke locomotive engine development program was initiated by Southwest Research Institute (SwRI) [17]
Brazil	1981	CESP-Energy Company of Sao Paulo started research on Methanol locomotives [18]
South Africa	1983	A consortium of South African universities, Sasol Alcohols, South African Department Agriculture and Fisheries, South African Transport Services started implementing a program towards oxygenated fuels for transportation by expanding Methanol/ethanol/propanol production [18]
USA	1989	Methanol Institute was founded as a non-profit organization. Its purpose was to lobby with the US Congress to support Methanol fuel markets
Canada	2008	Blue Fuel Energy-a Canadian company in British Columbia started to implement green Methanol production from biomass

1.4 Motivation

Diesel locomotives operate in extreme environmental conditions such as shocks, vibrations, dust etc. DME and Methanol should fulfill not only these operating requirements but also reliability requirements. There are two types of diesel locomotives used in India. The first is a 4-cycle engine, which has a design origin of over six decades ago (ALCO Locomotives) and the second is a more recent 2-cycle engine (GE EMD Locomotives), which was designed three decades ago. Therefore, a detailed design review and a comprehensive review along with feasibility of implementing DME and Methanol has to be investigated. Research and development of DME and Methanol fueled locomotives is necessary due to very attractive and relevant benefits they offer such as production from domestic resources such as high ash coal etc., clean combustion and emission characteristics, and relatively lower soot emissions. Hence it becomes essential to review technical feasibility of such new fuels for use in locomotives in Indian context.

2 Production of DME and Methanol

DME is conventionally produced from fossil resources rich in hydrocarbons. These include natural gas, coal, forest products, organic waste, energy crops, and agricultural by-products, which are converted to synthesis gas (syngas) via gasification. Syngas can then be used for the synthesis of Methanol, which can be further converted to DME. The MSW from cities can also be recycled and converted to syngas and finally to energy (Methanol and DME). This will improve the urban air quality as well as it will take care of large volumes of MSW generated in the urban settings.

2.1 Agricultural Residues

North India, especially Delhi area faces severe fog and pollution relates issues in November-December because of burning of the agricultural residues in huge swaths of fam lands in the states of Punjab and Haryana, at the end of the cultivation season. These agricultural residues can be used for production of syngas and then Methanol and DME. This will not only generate revenues for the farmers, but also, it will give huge respite to the residents of North India from several pollution issues and foggy days that they face every year.

Residue to product ratio (RPR) is defined as the amount of residue per kilogram of the product. Table 4 shows the utilization rate and RPR of different agricultural residue. Physico-chemical properties of various agricultural residues are shown in Table 5.

Table 4 Agricultural residue to product ratio (RPR) and utilization rate of each residue [19]

Agricultural products	Residue	RPR	Utilization rate %		
			Min	Max	Average
Sugar cane	Tops and leaves	0.20	10	30	20
	Bagasse	0.30	100	100	100
Rice in-season	Husk	0.23	70	80	75
	Straw	1.19	50	50	50
Rice off-season	Husk	0.23	70	80	75
	Straw	1.19	50	50	50
Cassava	Stem and leaves	0.12	60	80	70
	Rhizome	0.09	0	0	0
Maize	Stem and leaves	0.89	10	10	10
	Corncob	0.19	70	70	70
Palm	Fiber	0.15	100	100	100
	Empty fruits bunches	0.22	50	60	55
	Leaves	0.27	100	100	100
	Shell	0.13	70	80	75

Table 5 Physico-chemical properties of various agricultural residues (% w/w) [19]

Agri. product	Type of residue	Moisture	Volatile	Fixed	Ash	C	H	N	O	S	LHV (kJ/kg)
Sugarcane	Tops and leaves	9.2	68	16.9	6.1	41.6	5.1	0.4	37.4	0.2	15,479
Rice	Stem	10	61	18.9	10.4	38.2	5.0	0.6	35.3	0.1	9299
Maize	Stem and leaves	8.4	70	16.3	5.3	47.3	5.1	0.8	40.6	0.2	9830
Cassava	Rhizome	59	31	8.1	1.5	18.8	2.5	0.3	17.5	nt.	8225
Palm	EFB	58	30	8.0	2.9	15.1	1.5	2.6	19.1	0.1	7240

2.2 Production Process

DME can be produced by two methods; (a) direct method, and (b) dehydration reaction of Methanol called indirect method [20]. Figure 3 shows the schematic of both processes. A dual catalyst can allow Methanol synthesis and dehydration in one unit with no intermediate Methanol separation. The direct method has higher efficiency because Methanol itself is synthesized from the synthesis gas. Both the one-step and two-step processes are commercially available.

Figure 4 shows the schematic of the indirect process for production of DME from Methanol.

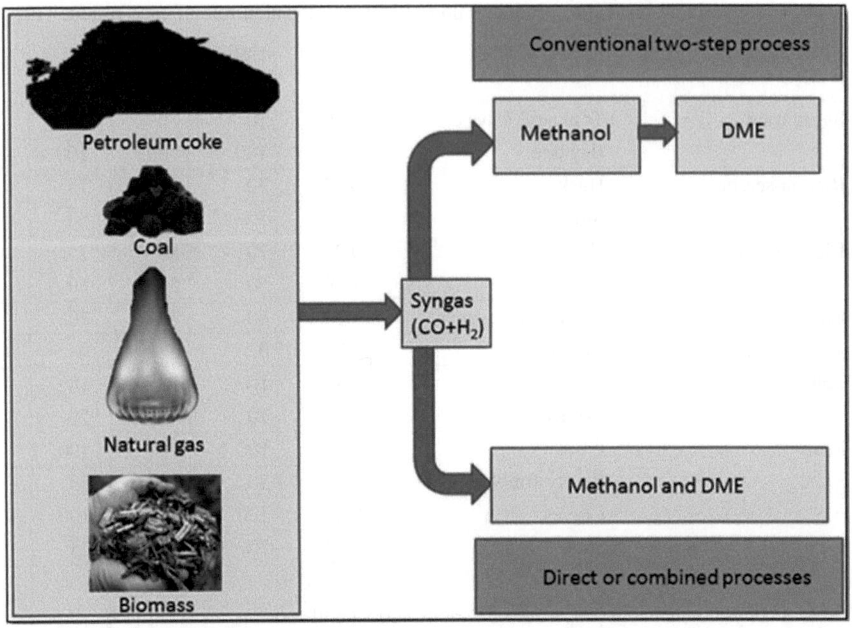

Fig. 3 Steps for DME and Methanol production

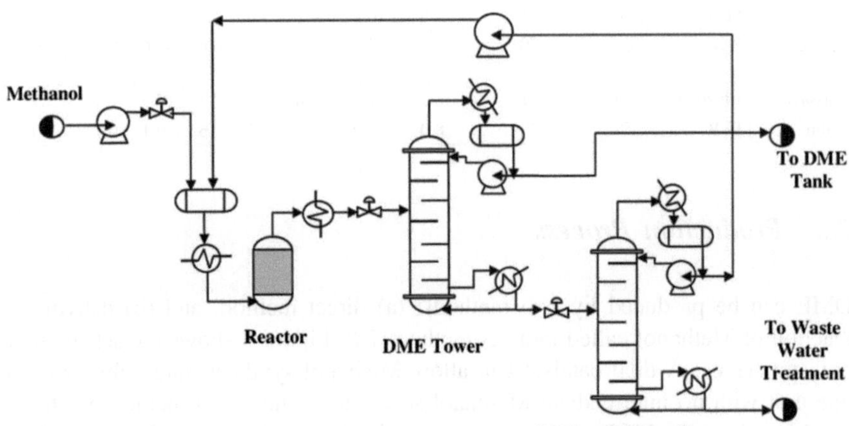

Fig. 4 Schematic of indirect synthesis process [21]

The commercialized process reaction of DME production from Methanol dehydration is given by reaction (1):

$$2CH_3OH \rightarrow CH_3OCH_3 + H_2O \quad (1)$$

The direct synthesis of DME from syngas containing H_2, CO and CO_2 follows mainly two overall reactions: Reaction (2) with water–gas shift reaction taken into account and reaction (3) without it.

$$3CO + 3H_2 \rightarrow CH_3OCH_3 + CO_2 \quad (2)$$

$$2CO + 4H_2 \rightarrow CH_3OCH_3 + H_2O \quad (3)$$

For direct production of DME, kinetic rate of reaction should be high. Therefore, syngas needs to be fed at a pressure of 5 MPa and temperature at 270 °C. This reaction is highly exothermic in nature and has to be handled with utmost precaution, in order to avoid any accident. Direct synthesis is preferred over indirect method because it results in minimum wastage of natural gas. However it is one of the most complex reactions of methane conversion. Through absorption, flash and distillation, H_2, N_2, CH_4, and CO_2 are removed, Methanol is recovered and then final product i.e. DME is obtained.

Methanol synthesis from syngas [22] consists of the three following reactions [23, 24]:

$$CO + 2H_2 \rightleftharpoons CH_3OH \quad (4)$$

$$CO_2 + 3H_2 \rightleftharpoons CH_3OH + H_2O \quad (5)$$

Reverse water gas shift

$$CO_2 + H_2 \rightleftharpoons CO + H_2O \quad (6)$$

All three reactions are reversible hence it is important to control the process conditions such as temperature, pressure and synthesis gas mixture. It can also be noted that reactions (4) and (5) are exothermic, i.e., the processes produce heat therefore they require cooling. Some heat is normally recovered and used for other parts of the synthesis. Reaction (6) describes reverse water-gas shift reaction, which produces carbon monoxide from carbon dioxide and hydrogen. The carbon monoxide then reacts with hydrogen to produce Methanol (Reaction 5). Reaction 5 is actually merely the sum of reactions (4) and (6) [3]. To synthesize Methanol, not only is a specific H_2/CO ratio of 2 in the synthesized gas is essential but also a $(H_2 - CO_2)/(CO + CO_2)$ ratio, called stoichiometric number, equal to or slightly above 2 is required [4].

3 Properties, Opportunities and Challenges of DME and Methanol

DME is the simplest ether, with a chemical formula of CH_3OCH_3. It is a volatile organic compound, however it is non-teratogenic, non-mutagenic and non-toxic. Methanol is the smallest primary alcohol, with a chemical formula of CH_3OH. Their use as an alternate fuel in locomotives offers a huge opportunity however there are many technical challenges as well.

3.1 Properties of DME and Methanol

DME has a boiling point of −25 °C and has a vapor pressure of about 0.6 MPa at ambient temperature, which enables it to be easily liquefied and stored in liquid state at a relatively lower pressures. Methanol on the other hand is a liquid fuel at low temperature and can be handled as any other conventional liquid fuel. Table 6 shows the typical fuel properties of DME and Methanol compared to baseline mineral diesel [25, 26].

Combustion properties: Auto-ignition temperature is one of the most important factors affecting combustion in a CI engine. Fuel-air mixture is heated during the ignition delay period in a CI engine. The fuel-air mixture gets self-ignited after attaining its auto ignition temperature. DME has lower auto-ignition temperature compared to baseline mineral diesel (DME: 508 K, mineral diesel: 523 K). However this auto-ignition temperature decreases with increase in in-cylinder

Table 6 Typical fuel properties of DME and Methanol [25, 26]

Property	Unit	Diesel	DME	Methanol
Chemical formula	–	–	$CH_3–O–CH_3$	CH_3OH
Molar mass	g/mol	170	46	32
Carbon content	% w/w	86	52.2	37.5
Hydrogen content	% w/w	14	13	12.5
Oxygen content	% w/w	0	34.8	50
Liquid density	kg/m^3	831	667	796
Cetane number	–	40–50	>55	5
Auto-ignition temperature	K	523	508	743
Stoichiometric air/fuel ratio	–	14.6	9.0	6.49
Boiling point/range at 1 atm	K	450–643	248.1	337.5
Lower heating value	mJ/kg	42.5	27.6	19.9
Ignition limits	$\% V/V_{air}$	0.6–6.5	3.4–18.6	6.7–36
Kinematic viscosity of liquid	cSt at 40 °C	3	<0.1	0.596
Surface tension (at 298 K)	N/m	0.027	0.012	22.18
Vapor pressure (at 298 K)	kPa	<10	530	32

pressure. Due to the higher cetane number of DME, auto ignition is relatively easier than mineral diesel.

High cetane number: Cetane number of DME is > 55, whereas the range of cetane number for mineral diesel is 40–51. DME has higher cetane number than mineral diesel, which leads to a shorter premixed combustion phase and shorter ignition delay. This dictates combustion, performance and emission characteristics of a CI engine. On the other hand, Methanol's cetane number (5) is significantly lower than diesel and DME.

Oxygenated fuel: Presence of oxygen in a fuel promotes combustion reactions thereby it reduces PM emissions. Complete absence of C–C bonds in DME results in smokeless combustion and soot-free emissions. The lower bond energy of C–O bonds compared to C–H bonds results in a shorter ignition delay, which reduces PM emissions further.

Lubricity and viscosity: DME does not have good lubricity property and is also less viscous compared to mineral diesel. This creates issues in fuel injection equipment (FIE) and corrodes vital parts of FIE and the engine. Affected components of FIE include components of fuel delivery system, plungers and fuel feed pump. Viscosity of DME is lower than the prescribed ASTM standard range for diesel (1.39–4.2 cSt at 40 °C). Therefore, additives or diesel or biodiesel has to be added to DME in order to improve its lubricity and viscosity characteristics so that it can become suitable as a fuel for CI engines.

Low heating value: Both DME and Methanol have lower heating values than mineral diesel due to presence of oxygen in their molecules. Therefore higher amounts of fuel is required to be injected in order to achieve torque equivalent to diesel fueled engines. This is a drawback of using such fuels for locomotives, where fuel consumption is quite high. This requires larger fuel storage tanks, and FIE systems, which can handle greater volume flow rates of fuel for the same power output and range of these locomotives.

Higher latent heat: DME and Methanol have relatively higher latent heat of vaporization compared to baseline mineral diesel. Hence they absorb higher heat during fuel evaporation process, thus reducing in-cylinder temperatures significantly compared to mineral diesel. This significantly cools the in-cylinder charge.

Low boiling point: DME spray emerging from an injector evaporates very quickly due to its lower boiling point compared to mineral diesel. This results in a temperature drop in premixed combustion zone during combustion, which reduces the formation of nitrogen oxide.

Compressibility characteristics of DME: The compressibility of DME is higher than that of mineral diesel. In a CI engine, compression work for DME in the high pressure fuel injection pump will be higher than that required for mineral diesel because DME has higher compressibility and lower density than mineral diesel.

3.2 Opportunities and Technical Challenges

Some of the opportunities offered by DME/Methanol are given below [25, 26]:

1. DME/Methanol can be produced from multiple sources and are multi-purpose fuels.
2. Engines fueled by DME produce ultra-low emissions, whereas engines fueled by Methanol produce relatively lower emissions than mineral diesel.
3. Lower CO_2 is produced by Methanol and DME fueled engines.
4. Overall noise and combustion noise is lesser than diesel engines.
5. DME and Methanol fueled engines offer superior thermal efficiency than mineral diesel fueled engines. This leads to higher fuel economy and higher well-to-wheel efficiency.
6. They have better ignition characteristics than mineral diesel.
7. Highest land use efficiency than any other cultivable renewable fuel.

Some of the technical challenges offered by DME/Methanol are given below [25, 26]:

1. *Lower combustion enthalpy*: The energy content of DME and Methanol is lower than mineral diesel therefore higher fuel quantities are required to be injected in every injection to provide energy equivalent to mineral diesel. This requires upsizing of FIE systems and related equipment.
2. *Lower viscosity and lubricity*: This results in harmful effects on the FIE such as corrosion and wear of moving components. DME may also cause high leakage from the FIE, which relies on small clearances for sealing.
3. Lower modulus of elasticity: The compressibility of DME is typically four to five times higher than mineral diesel. Therefore higher pumping work has to be done in case of DME engines.

Table 7 shows the opportunities and challenges of using Methanol and DME as an alternate fuel for locomotive traction.

Table 7 essentially outlines the changes required to be made in a locomotive engine, in order to make it compatible with the DME/Methanol.

3.3 Non-technical Challenges

In addition to technical challenges, there are few non-technical challenges/barriers, which need to be overcome before the DME and Methanol can be used as fuel in India in locomotives and other applications. Although DME and Methanol offer outstanding features as alternate fuels for CI engines, these alternative fuels have not been seriously explored in India. Table 8 shows the non-technical barriers to DME and Methanol usage in India.

Since there is no example of a locomotive engine operating on DME/Methanol, the next section discussed various technical aspects of DME/Methanol fueled

Table 7 Opportunities and challenges of using Methanol and DME for locomotive traction

Opportunities	Challenges
1. Can be used as a fuel for locomotives with minor modifications in FIE 2. Have large potential for railway traction market 3. Relatively higher cetane number 4. Quieter combustion 5. Clean burning characteristics: Soot-free combustion of DME and nearly smoke-less combustion for Methanol 6. 100% SO_x reduction 7. Relatively higher volatility than diesel 8. Existing LPG infrastructure can be used for transportation/fuel distribution	1. Technically sound knowledge to be developed in India for use in locomotives 2. Lower fuel lubricity requires lubricating agent to be added to the fuel 3. Lower viscosity causes higher fuel leakage from the FIE system 4. Government regulations to be put in place in India 5. Appropriate fuel storage system needs to be developed

Table 8 Barriers to DME and Methanol usage in India

Barriers to DME and Methanol usage
1. High capital investment required in Methanol and DME production plants
2. Lack of public awareness about the opportunities offered by these fuels
3. No end use for these fuels therefore there is no domestic production as of now
4. No active lobbying group, since this is not on the radar of big business entities

automotive engines, which are relatively smaller, but they offer significant clarity into combustion behavior of these fuels in CI engines. These findings can certainly be helpful in developing a DME/Methanol fueled locomotive engine.

4 DME and Methanol Fueled CI Engines

CI engines produce higher torque, higher thermal efficiency and lower HC and CO emissions compared to their SI counterparts. However CI engines suffer from issues such as relatively higher NO_x, smoke and particulate emissions. Emission legislations are becoming stringent with time globally. Therefore it is difficult for CI engines fueled by either diesel or biodiesels to comply with these ever-tightening emissions norms.

Methanol is explored in other parts of world however in India, it is yet to be implemented. Few Indian researchers have reviewed this alternative fuel, however experimental investigations are hardly reported in open literature. Indian automotive companies have focused on several other alternative fuels such as biodiesel, ethanol etc., however they have also completely ignored this option due to its limited availability. Major issue that needs to be addressed for use of DME/Methanol as an automotive/locomotive fuel in India, is to design a FIE system, which can overcome poor lubricity, lower viscosity, lower LHV and lower modulus of elasticity offerred by the target fuels. In order to improve lubricity, lubricating additives can be added to the fuel. However materials to improve lubricity and the heating value of DME/Methanol are yet to be developed in India. One of the characteristics of DME is its solubility in hydrocarbon fuels. Therefore blending of fuels is indispensable for DME engine applications in automotive/locomotive applications. Detailed investigations are required on the capabilities of various fuel injection strategies such as multiple injections, optimization of injection and spray angles, fuel injection pressure, etc., in order to achieve complete combustion with lower emissions and to gain control over the engine operating parameters. Use of newer innovative combustion technologies such as HCCI, PCCI, LTC, etc. also appears to have great potential in developing an optimized DME/Methanol fueled engine with controlled combustion and emissions.

During past few years, researchers have implemented numerous techniques/developed numerous technologies such as EGR, catalytic converters and retarding the injection timing etc. in order to reduce engine-out emissions. Few researchers suggested implementing enhanced swirl and squish to overcome problems of relatively higher HC and CO emissions. These techniques were helpful to a certain extent only. However in order to overcome majority of these issues, use of DME and Methanol could be promising out-of-the-box solution, which required only minor engine modifications. These fuels can be blended with diesel in different proportions so that relative benefits of each fuel could be exploited.

4.1 DME Engine Investigations

Due to relatively higher cetane number, DME combustion starts relatively earlier compared to mineral diesel. Addition of DME to diesel changes the physico-chemical properties of the fuel. As the DME fraction in diesel is increased, the density, kinematic viscosity and calorific value decreases however C/H ratio, oxygen content and cetane number increases [27, 28]. Use of 100% DME or Methanol as fuels in CI engines causes various issues due to their lower viscosity and lubricity. Major modifications in the FIE system are essential, if higher blends of DME or Methanol are used in CI engines. DME exists in gas phase at ambient conditions therefore it is required to be converted into liquid phase so that its volume reduces and it can be handled by the FIE system. Vehicle manufacturers including Isuzu, Nissan, Shanghai Diesel and Volvo have developed heavy vehicles running on DME fueled CI engines.

The fuel spray of DME is significantly different than mineral diesel because of its physical properties. In order to have good combustion and performance characteristics of DME fueled engine, it is essential to understand the spray behavior, which is fundamental to the fuel-air mixing process and hence combustion. Figure 5 shows

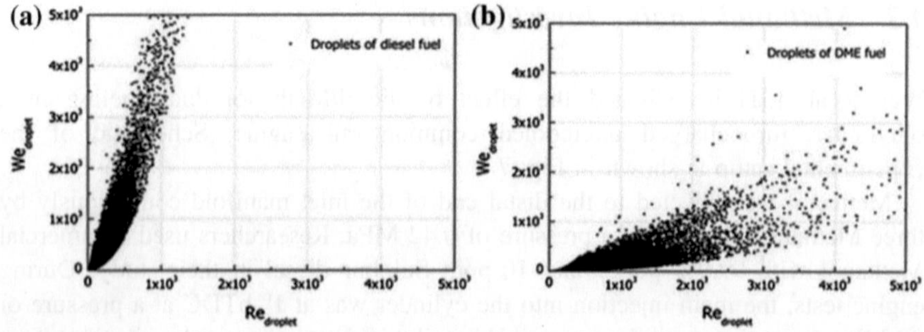

Fig. 5 Relation between Weber and Reynolds numbers for **a** diesel and **b** DME sprays [29]

the relationship between weber number and Reynolds number for diesel and DME. These dimensionless numbers were calculated from droplet diameter, velocity and surface tension. Diesel showed a sharp slope and DME showed a gentler slope because the rate of increase in Reynolds number was greater than that of Weber number. This was possibly due to lower viscosity of DME. Another point to be noted is that Reynolds number was responsible for spray breakup of DME and Weber number was the main breakup factor for mineral diesel.

Figure 6 shows combustion characteristics of diesel and DME with single injection in a CI engine. Park et al. [30] compared fuel injection timings at TDC and 30° bTDC. For both fuel injection timings, combustion of DME occurred before that of mineral diesel due to its relatively higher cetane number, which led to shorter ignition delay in case of DME.

As the fuel was injected at TDC, ignition of both test fuels (diesel and DME) occurred in the expansion stroke and the peak cylinder pressure of DME was higher than mineral diesel. The rate of heat release (ROHR) and rate of pressure rise (ROPR) peaks for DME were slightly lower than mineral diesel, when the fuel was injected at TDC. This was because of shorter ignition delay and insufficient fuel–air mixing time for DME. On the other hand, when the fuel was injected at 30° bTDC, ROHR and ROPR peaks for DME were relatively higher than mineral diesel due to longer injection duration.

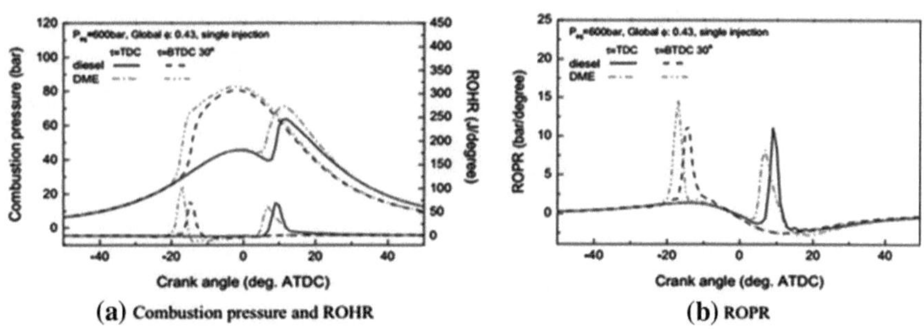

Fig. 6 Combustion characteristics of diesel and DME fueled engine [30]

4.2 Methanol Engine Investigations

Wei et al. [31] investigated the effect of diesel/Methanol dual fueling in a 6-cylinder, turbocharged intercooled common rail engine. Schematic of the experimental setup is shown in Fig. 7.

Methanol was injected to the distal end of the inlet manifold continuously by three Methanol injectors at a pressure of 0.42 MPa. Researchers used commercial Methanol with 99.9% purity and 10 ppm Sulphur diesel in their study. During engine tests, the main injection into the cylinder was at 1° bTDC at a pressure of 80 MPa. Operating conditions included no pilot injection (i.e. single injection), four different pilot-injection quantities from 2 to 8 mg, with pilot injection timing 18° bTDC and five different pilot injection timings from 6° bTDC to 18° bTDC with fixed pilot injection quantity of 6 mg.

Figure 8 shows the effect of pilot injection timing and fuel injection quantity on in-cylinder pressure, ROHR and mass burn fraction (MFB) for 50% Methanol and 50% diesel as test fuel. In case of single injection timing, peak in-cylinder pressure was significantly lower. As the pilot injection timing was advanced, peak in-cylinder pressure increased. On the other hand, as the pilot fuel injection quantity was increased, the peak in-cylinder pressure also increased. This was due to shorter ignition delay for main-injection. Advancement of pilot injection timing indicated that combustion duration was shorter and led to improvement in efficiency. The maximum value of ROHR for both pilot-injection and main-injection was almost same. MFB increased with increasing pilot-injection quantity.

There are numerous other studies on automotive engines using DME and Methanol as fuels. From these studies, it emerges that the main challenge for developing a DME or Methanol fueled locomotive will be to develop a FIE system, which will be able to deal with all the challenges that these fuels offer due to their very distinct physico-chemical properties.

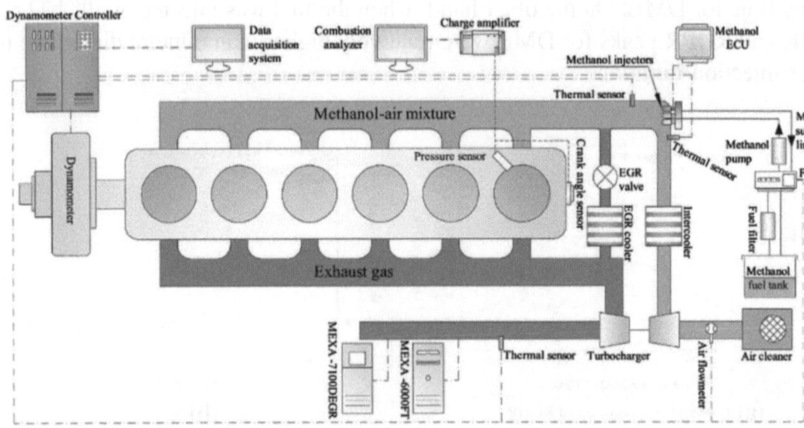

Fig. 7 Schematic of Methanol test engine [31]

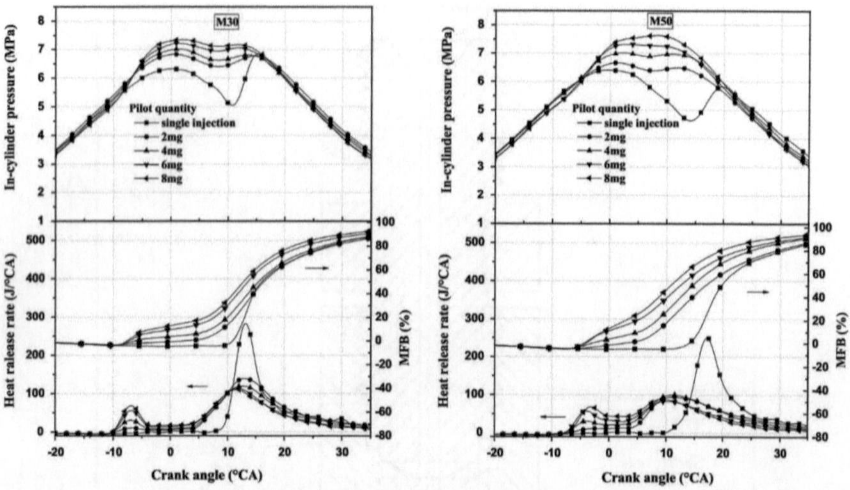

Fig. 8 Effect of pilot injection timing and pilot fuel quantity on in-cylinder pressure, ROHR and MFB [31]

5 Modifications in FIE for DME/Methanol Locomotive Development

DME is a gas at atmospheric pressure and it liquefies with modest pressurization. In order to liquefy gaseous DME, it is necessary to exert pressure greater than 5.1 bar at 25 °C [25, 26]. DME also has very low viscosity, leading to high internal leakage in supply pumps, solenoid valves, and fuel injectors. Such limitations demand significant modifications in the conventional FIE to enable use of DME in a diesel engine. DME lacks lubricity therefore it leaks from very small clearances in the seals. In an experiment in HFRR, addition of 100 ppm fatty acid based lubricity improver in DME reduced the wear scar diameter by 400 μm. It was found that rapeseed oil methyl ester, diesel, vegetable oil and Castrol oil are good DME lubricity improvers [32, 33].

The leakage rate of DME from the fuel injectors due to its lower viscosity and lower lubricity could be as high as 50%. DME/Methanol also attack the materials, which are prone to corrosion. Therefore conventional sealing material in the fuel injectors should be replaced by anti-corrosive materials. Wear in the moving components of fuel injector can be addressed by adding lubricity additives in the DME. Therefore it is important to develop a leakage-free, durable FIE system for DME for use in locomotive engines.

DME has very low boiling point therefore it becomes important to pressurize DME so that it remains liquid under all operating and environmental conditions and also in order to prevent cavitation in the fuel injection nozzle, which may lead to unstable fuel injection. Because of lower energy density of DME, injector needs to

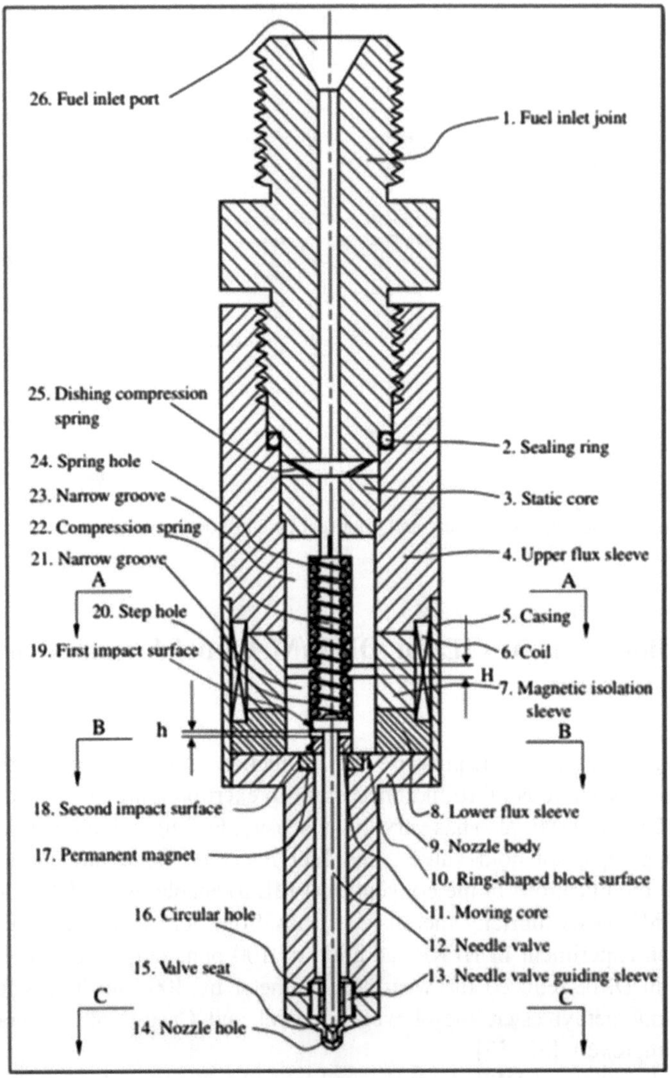

Fig. 9 Sectional view of novel common-rail type DME injector [34]

inject higher volume of the test fuel. Therefore it needs to be modified in order to provide necessary fuel delivery capacity in terms of higher volume. The nozzle needle opening pressure must be adjusted because of differences in fuel density, modulus of compressibility and heating value of DME. The needle opening pressure could be considerably lower than the pressure required for mineral diesel operation.

In electromagnetic fuel injectors, electromagnetic forces are directly used to drive the needle valve. However with DME as test fuel, injector needs large electromagnetic force to drive the needle valve. Therefore there exist a slower response for opening and closing the injector. Xu et al. [34] worked on the development of dedicated novel common-rail type DME injector. This design will not have any plunger matching parts. Therefore this design will not have any sealing or wear related problems. Oil return is absent from the injector, which can be directly driven by 24 V power supply. This typical design is shown in Fig. 9.

The fuel tank needs to be pressurized upto 8–10 bar so that DME remains in liquid state. This pressure also forces the fuel to flow through the FIE system, thus eliminating the need for fuel transfer pump. Fuel is further pressurized to a pressure of 10–12 bar through a gear pump so that DME remains in liquid state. Thereafter the DME in liquid state is further pressurized by high pressure fuel pump, before injection into the locomotive engine cylinder. DME/Methanol have a tendency to attack the rubber parts therefore the O-rings used in the FIE system, including storage tanks should be tested frequently to ensure leak-proof system. There also exists a fuel return line, which should be maintained at a pressure of 10–12 bar, depending on the pressure setting of the back pressure regulator.

6 Safety of Methanol and DME Fueled Locomotives

Major accidents occurring in IR are of the following forms [3]:

(i) Level crossing accidents
(ii) Derailments
(iii) Fire
(iv) Collisions.

These accidents are independent of the type of fuel used. Methanol will not have any additional major safety issues. However, DME may add its own complications. DME has low viscosity and there is a high chance of leakage through the fuel injection system. Similar to natural gas, DME burns with visible flame. This shows that DME is not that unsafe fuel. However, there is a requirement of a proper ventilation system in the locomotive because DME vapors are heavier than air. Heavier molecules of DME will have tendency to settle down towards lower ground, which may possibly be hazardous. DME is a volatile organic compound, which is non-carcinogenic, non-toxic, non-teratogenic, and non-mutagenic. Although DME has shown higher flammability but it must be handled with extreme care. It is considered as a stable compound and it does not decompose to peroxides in ambient conditions. DME is found to be soluble in both polar and non-polar solvents, which means that it is partially miscible in water.

Comparison of the global warming potential (GWP) of various hydrocarbons reveals that DME is atmospherically benign and does not contribute significantly to the global warming. GWP is a relative measure of the quantity of heat, a greenhouse gas traps in the atmosphere. It compares the amount of heat trapped by a certain mass of gas to the amount of heat trapped by a similar mass of CO_2. Precisely, it is a measure of how much energy the emissions of 1 ton of a gas will absorb over a given period of time, relative to the emissions of 1 ton of CO_2. The larger the GWP, the more that a given gas warms the earth compared to CO_2 over that time period. The time period usually used for GWPs is 100 years. Typical GWP value of DME is 1 and Methanol is 2.8 for 100 years [35, 36].

7 Summary

Alternative fuel options for locomotives in India	• IR is witnessing an exponential increase in fuel bill due to increased tariffs, consumption and increased number of locomotives • There is a growing concern for the environmental pollution from locomotives in India. India is seeking non-petroleum alternative fuels for locomotives • Domestic availability, economics, acceptability, emissions compliance, national security, technology, versatility could be the criteria to judge new alternative fuels for rail road traction • Two new alternative fuels have been proposed for locomotives. Amongst synthetic energy resources, DME can be most appropriate solution in near future. Methanol is an oxygenated fuel and it helps reduce emissions in a CI engine. It can be blended with diesel for use in locomotives
Production of DME and Methanol	• MSW can be recycled and converted to energy. This will improve urban air quality as well as energy security • DME can be produced by two methods; (a) direct method and (b) indirect method involving dehydration of Methanol • Methanol can be produced from syngas, which can be produced from variety of feedstock such as high ash coal, agro-residues and MSW
Combustion characteristics of Methanol and DME	• For same energy input, Methanol and DME engine shows better engine performance and emissions characteristics compared to diesel fueled engine
Emissions characteristics of Methanol and DME	• DME engine shows almost zero soot emission but slightly higher level of NO_x compared to diesel fueled engine • The CO and HC emissions are reduced remarkably compared to conventional diesel engine • In order to reduce the NO_x emissions, EGR or catalytic converters can be used

(continued)

(continued)

Driving characteristics of Methanol and DME fueled locomotives	• In order to evaluate driving performance of Methanol and DME fueled locomotives, endurance test and driving tests of real tracks need to be conducted
Safety of Methanol and DME Fueled Locomotives	• DME has lower viscosity and there are higher chances of leakage through the FIE system • Similar to natural gas, DME burns with visible flames • DME shows no signs of carcinogenicity, and no evidence of mutagenicity or teratogenicity • Unlike DME, Methanol is a highly volatile, flammable, colorless, and toxic liquid. Inhalation and skin absorption of Methanol may lead to toxic effects such as headache, blindness or serious visual impairment and even death in case of serious exposure

8 Action Plan for Developing DME/Methanol Fueled Locomotives

In India, biomass-to-Methanol/DME is foreseen to be the most energy-efficient pathway for locomotive fueling in near future. However, there are several technological interventions and exhaustive experiments, which need to be conducted, before such a locomotive can be developed and put to use.

These specific action points are:

- Fuel injection system development
- Combustion optimization
- Emission development and certification
- Component refinement and cost reduction
- Locomotive trials and optimization of systems and components
- Reliability and safety demonstration
- Fuel Infrastructure development
- Fuel certification
- Infrastructure creation for fuel production and distribution
- Process refinement: efficiency improvement and cost reduction.

IR requires finances for modernization however the required budgetary support is dodgy as on today. DME and Methanol certainly have potential to be used as fuel in locomotives, either as 100% replacement of diesel or as partial replacement, preferably by blending them with diesel in certain proportions to begin with. Unless government gives an aggressive push to these fuels, IR will continue using conventional fuels. In this context, a long-term plan must be proposed by IR and the policy makers of India should champion it. The key area of research involves development and durability of fuel-injection system and its integration with the existing fleet of locomotives. These DME/Methanol locomotives will partly substitute existing locomotives or can be used to create additional capacity in the IR network.

References

1. http://indianrailways.gov.in/railwayboard/uploads/directorate/stat_econ/IRSP_2013-14/pdf/Statistical_Summary/Summary%20Sheet_Eng.pdf. Accessed on 12th November 2016
2. Railways I. Year book 2011. Indian Railways, New Delhi. 2011
3. Indian Railways, Life line of nation, budget 2015–2016. http://www.indianrailways.gov.in/railwayboard/uploads/directorate/finance_budget/Budget_2015-16/White_Paper-_English.pdf. Accessed on 12th November 2016
4. Casadei S, Maggioni A (2016) Performance testing of a locomotive engine after treatment pre-prototype in a passenger cars chassis dynamometer laboratory. Transp Res Procedia 31 (14):605–614
5. Youn IM, Park SH, Roh HG, Lee CS (2011) Investigation on the fuel spray and emission reduction characteristics for dimethyl ether (DME) fueled multi-cylinder diesel engine with common-rail injection system. Fuel Process Technol 92(7):1280–1287
6. Park SH, Lee CS (2013) Combustion performance and emission reduction characteristics of automotive DME engine system. Prog Energy Combust Sci 39(1):147–168
7. Crookes RJ, Bob-Manuel KD (2007) RME or DME: a preferred alternative fuel option for future diesel engine operation. Energy Convers Manag 30, 48(11):2971–2977
8. Junjun Z, Xinqi Q, Zhen W, Bin G, Zhen H (2008) Experimental investigation of low-temperature combustion (LTC) in an engine fueled with dimethyl ether (DME). Energy Fuels 23(1):170–174
9. Technical information, Toxicity summary for dimethyl either (DME) Dymel A Propellant, http://www.aboutdme.org/EFIClient/files/ccLibraryFiles/Filename/000000000640/DME_Technical_-_Toxicity_Summary.pdf. Accessed on 12th November 2016
10. Li Y, Deng D, Xing X, Chen N, Liu X, Xiao X, Wang Y (2016) A high performance Methanol gas sensor based on palladium–platinum–In $2O_3$ composited nanocrystalline SnO_2. Sens Actuators B: Chem 31(237):133–141
11. Liu Q, Kirchhoff JR (2007) Amperometric detection of Methanol with a Methanol dehydrogenase modified electrode sensor. J. Electroanal Chem 15, 601(1):125–131
12. Paine AJ, Dayan AD (2001) Defining a tolerable concentration of Methanol in alcoholic drinks. Hum Exp Toxicol. 20(11):563–568
13. Wells KC, Millet DB, Hu L, Cady-Pereira KE, Xiao Y, Shephard MW, Clerbaux CL, Clarisse L, Coheur PF, Apel EC, Gouw JD (2012) Tropospheric Methanol observations from space: retrieval evaluation and constraints on the seasonality of biogenic emissions. Atmos Chem Phys 12, 12(13):5897–5912
14. Kwanchareon P, Luengnaruemitchai A, Jai-In S (2007) Solubility of a diesel–biodiesel–ethanol blend, its fuel properties, and its emission characteristics from diesel engine. Fuel 31, 86(7):1053–1061
15. Jiao W, Wang Y, Li X, Xu C, Liu Y, Zhang Q (2016) Stabilization performance of Methanol-diesel emulsified fuel prepared using an impinging stream-rotating packed bed. Renew Energy 31(85):573–579
16. Dincer I, Hogerwaard J, Zamfirescu C (2015) Clean rail transportation options. Springer
17. Cataldi CR (1988) Broadened diesel fuel specifications for rail applications. American Society of Mechanical Engineers, New York, NY

18. Onion G, Bodo LB (1983) Oxygenate fuels for diesel engines: a survey of world-wide activities. Biomass 31, 3(2):77–133
19. Lecksiwilai N, Gheewala SH, Sagisaka M, Yamaguchi K (2016) Net energy ratio and life cycle greenhouse gases (GHG) assessment of bio-dimethyl ether (DME) produced from various agricultural residues in Thailand. J Clean Prod 134:523–531
20. Ogawa T, Inoue N, Shikada T, Ohno Y (2003) Direct dimethyl ether synthesis. J Nat Gas Chem 12(4)
21. Azizi Z, Rezaeimanesh M, Tohidian T, Rahimpour MR (2014) Dimethyl ether: a review of technologies and production challenges. Chem Eng Process 31(82):150–172
22. Aasberg-Petersen K, Nielsen CS, Dybkjær I, Perregaard J (2008) Large scale Methanol production from natural gas. Haldor Topsoe 22
23. Olah GA (2005) Beyond oil and gas: the methanol economy. Angewandte Chemie International Edition 44(18):2636–2639
24. Lange JP (2001) Methanol synthesis: a short review of technology improvements. Catal Today 1, 64(1):3–8
25. Park SH, Lee CS (2013) Combustion performance and emission reduction characteristics of automotive DME engine system. Prog Energy Combust Sci 28, 39(1):147–168
26. Arcoumanis C, Bae C, Crookes R, Kinoshita E (2008) The potential of di-methyl ether (DME) as an alternative fuel for compression-ignition engines: a review. Fuel 30, 87(7):1014–1030
27. Patil KR, Thipse DS (2012) The potential of DME-diesel blends as an alternative fuel for CI engines. Int J Emer Technol Adv Eng, 2250–2459
28. Ying W, Longbao Z, Hewu W (2006) Diesel emission improvements by the use of oxygenated DME/diesel blend fuels. Atmos Environ 30, 40(13):2313–2320
29. Kim HJ, Park SH, Suh HK, Lee CS (2009) Atomization and evaporation characteristics of biodiesel and dimethyl ether compared to diesel fuel in a high-pressure injection system. Energy Fuels 2, 23(3):1734–1742
30. Park SH, Yoon SH (2015) Injection strategy for simultaneous reduction of NOx and soot emissions using two-stage injection in DME fueled engine. Appl Energy 1(143):262–270
31. Wei H, Yao C, Pan W, Han G, Dou Z, Wu T, Liu M, Wang B, Gao J, Chen C, Shi J (2017) Experimental investigations of the effects of pilot injection on combustion and gaseous emission characteristics of diesel/Methanol dual fuel engine. Fuel 15(188):427–441
32. Zhuoyong H, Tiegang H, Longbao Z, Shenghua L, Hongyi D (2005) Study on lubricity improver of dimethyl ether as the fuel of compression ignition engine. J Xi'an Jiaotong Univ 3:003
33. Ramadhas AS (ed) (2016) Alternative fuels for transportation. CRC Press
34. Xu S, Wang Y, Zhang X, Zhen X, Tao C (2012) Development of a novel common-rail type dimethyl ether (DME) injector. Appl Energy 30(94):1–2
35. United nation frame work convention on climate change, compilation of technical information on the new greenhouse gases and groups of gases included in the fourth assessment report of the intergovernmental panel on climate change. http://unfccc.int/national_reports/annex_i_ghg_inventories/items/4624.php. Accessed on 12th November 2016
36. Change IP. Climate change 2007: The physical science basis. Agenda. 2007 May 31;6 (07):333.

Part IV
Locomotive Emission Reduction and Measurement

Exhaust After Treatment System for Diesel Locomotive Engines—A Review

Prashant R. Daggolu, Dinesh Kumar Gogia and T.A. Siddiquie

Abstract Indian Railways (IR) operates a large fleet of Diesel Locomotives. In recent times, a concerted effort is being made to understand the emission implications on the environment from these locomotives and its remedial measures. This article reviews the global standards for diesel locomotive emission norms and available after treatment solutions for diesel locomotives. Currently, there are no emission norms defined in India for locomotives and may be considered in future. The Diesel after treatment solutions include Diesel Oxidation Catalysts (DOC), Diesel Particulate Filter (DPF) and Selective Catalytic Reduction (SCR) systems. Initially, it may be advisable to employ only DOC solution on diesel locomotives of IR followed by a thorough study of emissions on representative routes. Such a study would yield data for a design of a complete after treatment solution.

1 Introduction

Indian Railways (IR) caters to extensive rail network in India, which is fourth largest in the world with 8.397 billion passengers and over 106 million tons of freight transported annually as of 2014 [1].

IR operates largely Electric and Diesel Locomotives with a minor share of steam locomotives in special roles. The roiling stock of locomotives of IR stood at Electric—5016; Diesel—5714; Steam—43 [2]. Apart from this, 70% of all traction in India is powered by Diesel [3]. This results in high diesel footprint across the network and hence consequent emissions.

So far, there has been no implementation of emission reduction measures or defining regulatory mechanisms for emissions from Diesel locomotives [3]. However, Research Designs & Standards Organization (RDSO), IR, Lucknow has taken up work in this regard. The efforts so far have been largely based on engine

P.R. Daggolu (✉) · D.K. Gogia · T.A. Siddiquie
Sud Chemie India Pvt. Ltd., A-1/2 Nandesari Industrial Area, Nandesari, Vadodara 391340, Gujarat, India
e-mail: Prashant.daggolu@sud-chemie-india.com

A.K. Agarwal et al. (eds.), *Locomotives and Rail Road Transportation*,
DOI 10.1007/978-981-10-3788-7_8

modifications such as use of high boost pressure turbochargers, higher compression ratio pistons and common rail electronic fuel injection systems. These steps result in reduction in emissions as well as they enhance the fuel efficiency.

2 Indian Diesel Locomotives

Diesel locomotives are generally driven by 6, 12, 16 cylinder engine (usually 16 cyl). In India these are manufactured by Diesel Locomotive Works, Varanasi. However several of the earlier locomotives have been imported or built under Technology Transfer (ToT) agreements in India. Most of the current Diesel locomotives are in the range of 3100–4500 HP with 16 cylinder engine. For example, the locomotive WDP 4D (see Fig. 1) is an indigenized version of EMD GT46PAC locomotive from General Motors-Electromotive Division. This a 4500 HP engine, 16 cyl, 182.6 L, operated by IR for hauling long range passenger trains.

3 Diesel Emissions

Diesel engines emit certain toxic emissions. There are 4 main components of emissions—Carbon Monoxide (CO), Unburnt Hydrocarbon (HC), Nitrogen Oxides (NOx) and Particulate matter (PM) from diesel engines.

Fig. 1 WDP-4D locomotive of Indian Railways

Table 1 Locomotive pollution in connecticut, USA

Rail line	NOx (tons per year)	PM (tons per year)	CO (tons per year)	HC (tons per year)
Commuter lines				
Shoreline east	88.2	3.1	9.0	4.3
New Haven line	360.7	11.3	37.3	19.1
Amtrak	868.2	26.0	110.2	37.1
Line haul				
CSX	70.5	1.8	7.2	3.5
Providence and Worcester	98.8	2.5	10.1	4.8
Total state-wide emissions	1486.4	44.7	173.8	68.7

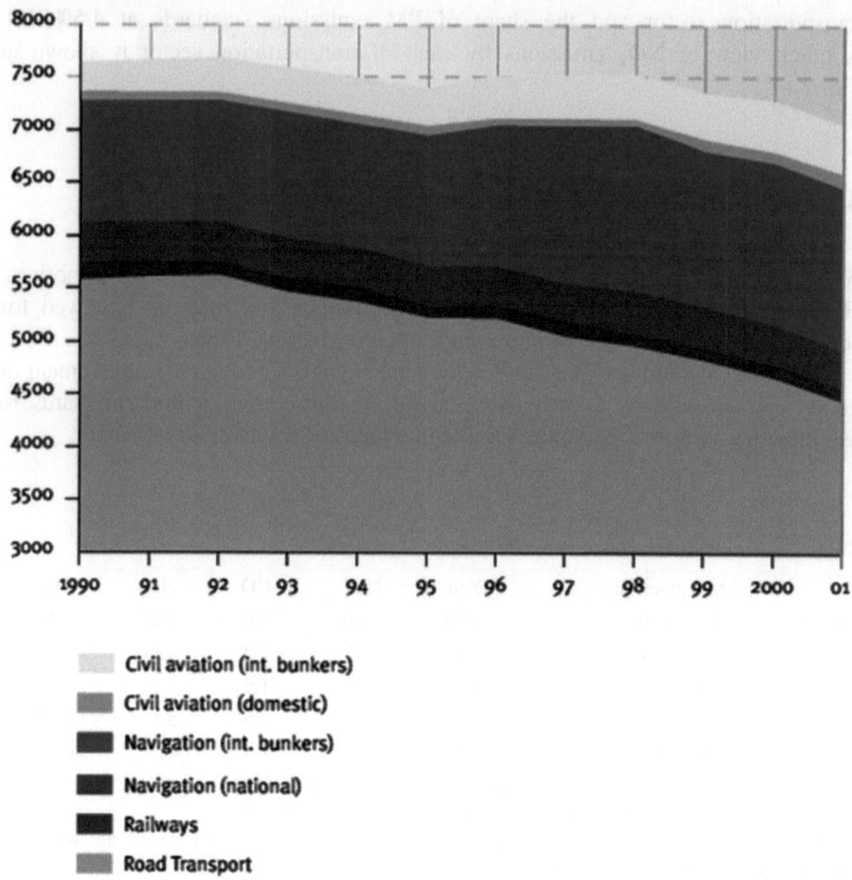

Fig. 2 The trend of NO_x emission of various transportation sector segments in Europe between 1990 and 2001

CO is a result of incomplete combustion in engine. CO is a highly toxic substance which has neurological implications in humans and its high concentrations can be fatal. Unburnt hydrocarbon can be carcinogenic too. NO_x is also known to be fatal at high concentrations. Even at low concentrations, it can affect the central nervous system of humans. PM is small solid particles in nm size range. They deposit in the respiratory system of humans and can lead to respiratory ailments on long-term exposure. In India, data of overall air pollution caused by diesel locomotives alone is not known. However, a study made in USA in the state of Connecticut has shown the data listed in Table 1.

One noteworthy point is that, among all diesel engines, locomotive diesel engines are considered to be the least polluting per hp output and have high efficiency, being large bore medium speed engines.

It should also be noted that in comparison to other modes of transportation, diesel emissions from locomotives is only a small proportion [4] of 1–3%. In Europe, NO_x emissions constitute about 3% of overall NO_x emissions from transportation sector and the share of PM emissions standards at 4.5% [5]. A quick view of NO_x emissions by each of transportation sector is shown in Fig. 2 [4].

4 Comparative Global Locomotive Emission Norms

As noted earlier, presently there are no emission norms in India for Locomotives. However, such norms exist in advanced economies and may be reviewed for comparison. The Locomotive norms in USA are given in Tables 2 and 3.

Line Haul locomotives are those, which are engaged in long range movement of goods and passengers. Switch locomotives are those used within rail yards to assemble trains to make them ready for line haul locomotives to take over.

Table 2 Line-haul locomotive emission standards, g/bhp-hr

	Manufacturing year	Date	HC	CO	NO_x	PM
Tier 0[a]	1973–1992[c]	2010	1.00	5.0	8.0	0.22
Tier 1[a]	1993[c]–2004	2010	0.55	2.2	7.4	0.22
Tier 2[a]	2005–2011	2010	0.30	1.5	5.5	0.10
Tier 3[b]	2012–2014	2012	0.30	1.5	5.5	0.10
Tier 4	2015 or later	2015	0.14[d]	1.5	1.3[d]	0.03

[a]Tier 0–2 line-haul locomotives must also meet switch standards of the same tier
[b]Tier 3 line-haul locomotives must also meet Tier 2 switch standards
[c]1993–2001 locomotive that were not equipped with an intake air coolant system are subject to Tier 0 rather than Tier 1 standards
[d]Manufacturers may elect to meet a combined NO_x + HC standard of 1.4 g/bhp-hr

Table 3 Switch locomotive emission standards, g/bhp-hr

	Manufacturing year	Date	HC	CO	NO_x	PM
Tier 0	1973–2001	2010	2.10	8.0	11.8	0.26
Tier 1[a]	2002–2004	2010	1.20	2.5	11.0	0.26
Tier 2[a]	2005–2010	2010	0.60	2.4	8.1	0.13
Tier 3	2011–2014	2011	0.60	2.4	5.0	0.10
Tier 4	2015 or later	2015	0.14[b]	2.4	1.3[b]	0.03

[a]Tier 1–2 switch locomotives must also meet line-haul standards of the same tier
[b]Manufacturers may elect to meet a combined NO_x + HC standard of 1.3 g/bhp-hr

In Europe, the locomotive emission norms have been combined with Off-road engine norms. Locomotives have to comply with such norms of appropriate engine power ratings. The norms can be seen in Table 4.

The locomotive norms would fall under category 7 (NRE-v/c-7). Here there is no distinction between the application of locomotives, in contrast to norms in USA.

In USA, the testing cycle is based on various weighting factors tested at different locomotive notches. The weighting factors are different for long haul and switch locomotives. The details are given below for the weighting factors [6] (Tables 5, 6, 7 and 8).

In case the locomotive is not equipped with multiple Idle conditions, then the following weighting factors are to be used.

The power ratings of the notches are to be as per the below table. If not, a different cycle has to be proposed, meeting the requirements of EPA. However, EPA suggests that the power ratings be complied with as close to the standard requirements as possible.

In Europe, however, the distinction between the long haul and switching locomotives are not made. The testing cycle is ISO 8178 F. The details of weighting factors of ISO 8178 cycles are given below [7].

For the rest of the world, locomotive emissions remain largely unregulated, including in India.

Table 4 Proposed Stage V emission standards for non-road engines in EU

Category	Ign.	Net power	Implementation date	CO	HC	NO_x	PM	PN
		kW		g/kWh				1/kWh
NRE-v/c-1	CI	$P < 8$	2019	8.00	7.50[a,c]		0.40[b]	–
NRE-v/c-2	CI	$8 \leq P < 19$	2019	6.60	7.50[a,c]		0.40	–
NRE-v/c-3	CI	$19 \leq P < 37$	2019	5.00	4.70[a,c]		0.015	1×10^{12}
NRE-v/c-4	CI	$37 \leq P < 56$	2019	5.00	4.70[a,c]		0.015	1×10^{12}
NRE-v/c-5	All	$56 \leq P < 130$	2020	5.00	0.19[c]	0.40	0.015	1×10^{12}
NRE-v/c-6	All	$130 \leq P \leq 560$	2019	3.50	0.19[c]	0.40	0.015	1×10^{12}
NRE-v/c-7 (Locomotives)	All	$P > 560$	2019	3.50	0.19[d]	3.50	0.045	–

[a] $HC + NO_x$
[b] 0.60 for hand-startable, air-cooled direct injection engines
[c] A = 1.10 for gas engines
[d] A = 6.00 for gas engines

Table 5 Standard duty-cycle weighting factors for calculating emission rates for locomotives with multiple idle settings

Notch setting	Test mode	Line-haul weighting factors	Line-haul weighting factors (no dynamic brake)	Switch weighting factors
Low idle	A	0.190	0.190	0.299
Normal idle	B	0.190	0.315	0.299
Dynamic brake	C	0.125	([a])	0.000
Notch 1	1	0.065	0.065	0.124
Notch 2	2	0.065	0.065	0.123
Notch 3	3	0.052	0.052	0.058
Notch 4	4	0.044	0.044	0.036
Notch 5	5	0.038	0.038	0.036
Notch 6	6	0.039	0.039	0.015
Notch 7	7	0.030	0.030	0.002
Notch 8	8	0.162	0.162	0.008

[a]Not applicable

Table 6 Standard duty-cycle weighting factors for calculating emission rates for locomotives with a single idle setting

Notch setting	Test mode	Line-haul	Line-haul (no dynamic brake)	Switch
Normal idle	A	0.380	0.505	0.598
Dynamic brake	C	0.125	([a])	0.000
Notch 1	1	0.065	0.065	0.124
Notch 2	2	0.065	0.065	0.123
Notch 3	3	0.052	0.052	0.058
Notch 4	4	0.044	0.044	0.036
Notch 5	5	0.038	0.038	0.036
Notch 6	6	0.039	0.039	0.015
Notch 7	7	0.030	0.030	0.002
Notch 8	8	0.162	0.162	0.008

[a]Not applicable

Table 7 Standard notch power levels expressed as a percentage of rated power

	Percent
Normal idle	0.00
Dynamic brake	0.00
Notch 1	4.50
Notch 2	11.50
Notch 3	23.50
Notch 4	35.00
Notch 5	48.50
Notch 6	64.00
Notch 7	85.00
Notch 8	100.00

Table 8 Weighting factors of B-Type ISO 8178 test cycles

Mode number	1	2	3	4	5	6	7	8	9	10	11
Torque, %	100	75	50	25	10	100	75	50	25	10	0
Speed	Rated speed					Intermediate speed					Low idle
Off-road vehicles											
Type C1	0.15	0.15	0.15	–	0.10	0.10	0.10	0.10	–	–	0.15
Type C2	–	–	–	0.06	–	0.02	0.05	0.32	0.30	0.10	0.15
Constant speed											
Type D1	0.30	0.50	0.20	–	–	–	–	–	–	–	–
Type D2	0.05	0.25	0.30	0.30	0.10	–	–	–	–	–	–
Locomotives											
Type F	0.25	–	–	–	–	–	–	0.15	–	–	0.60
Utility, lawn and garden											
Type G1	–	–	–	–	–	0.09	0.20	0.29	0.30	0.07	0.05
Type G2	0.09	0.20	0.29	0.30	0.07	–	–	–	–	–	0.05
Type G3	0.90	–	–	–	–	–	–	–	–	–	0.10
Marine application											
Type E1	0.08	0.11	–	–	–	–	0.19	0.32	–	–	0.30
Type E2	0.20	0.50	0.15	0.15	–	–	–	–	–	–	–
Marine application propeller law											
Type E3, mode #	1		2		3		4				
Power, %	100		75		50		25				
Speed, %	100		91		80		63				
Weighting factor	0.2		0.5		0.15		0.15				
Type E4, Mode #	1		2		3		4		5		

(continued)

Table 8 (continued)

Mode number	1	2	3	4	5	6	7	8	9	10	11
Power, %	100		80		60		40		0		
Speed, %	100		71.6		46.5		25.3		Idle		
Weighting factor	0.06		0.14		0.15		0.25		0.4		
Type E5, Mode #	1		2		3		4		5		
Power, %	100		75		50		25		0		
Speed, %	100		91		80		63		Idle		
Weighting factor	0.08		0.13		0.17		0.32		0.3		

Notes

- Engine torque is expressed in percent of the maximum available torque at a given engine speed
- Rated speed is the speed at which the manufacturer specifies the rated engine power
- Intermediate speed is the speed corresponding to the peak engine torque

5 Diesel After-treatment Systems

Diesel after-treatment systems mainly consist for 3 devices—(1) Diesel Oxidation Catalysts (DOC), (2) Diesel Particulate Filter (DPF), and (3) Selective Catalytic Reduction (SCR).

6 Diesel Oxidation Catalysts

The main component of a Diesel Exhaust after-treatment system shall be DOC. These are designed to remove toxic CO, HC and SOF. DOC is a catalytic coating (Wash-coat) made on monolithic substrates. The wash-coat is impregnated with precious metals, typically, Platinum (Pt), Palladium (Pd) or both Pt and Pd. DOC conduct oxidation reactions with O_2, which is available in abundance in Diesel engine exhaust. The reactions schemes are as below:

$$CO + 1/2O_2 \rightarrow CO_2 \tag{1}$$

$$HC + O_2 \rightarrow CO_2 + H_2O \tag{2}$$

$$SOF + O_2 \rightarrow CO_2 + H_2O \tag{3}$$

A schematic of DOC is shown in Fig. 3.

Fig. 3 Schematic of a diesel oxidation catalysts

7 Diesel Particulate Filter

As noted above, Diesel Particulate Matter has 3 components—Soot, SOF and Sulfates. Of these, soot requires a filter for removal. Filters are available in 2 forms:

i. Partial Flow Filters (PFF)
ii. Wall Flow Filters or Diesel Particulate Filters (DPF)

Partial flow filters as the name suggests are aimed at partial reduction of soot from exhaust (usually to a maximum of 50%). PFFs are usually metal substrates with foam separating the channels. The foam is a proprietary material from the PFF manufacturers. When the exhaust passes through the foam, the soot particles are trapped in the foam and hence filtered out. However, since the channels are open at the other end, not all soot particles flow through the foam and filter out. Hence this is a partial filtration technique.

DPFs on the other hand, result in >95% filtration. These are ceramic honeycomb structures with alternate blocked channels on each side. i.e. the channel which has entry open will have the exit closed. This forces the entire exhaust to flow through the wall. The walls are made by special high porosity material. These allow gaseous exhaust to flow through but traps the soot.

It has to be also understood that the DPF may be used as it is or can be catalytically coated. If coated then it is referred as 'catalyzed diesel particulate filter' or cDPF. A Schematic of such a system is shown in Fig. 4.

The filters have finite capacity of soot storage and have to be periodically regenerated. The regeneration is done in 2 ways:

i. Active regeneration
ii. Passive regeneration

Soot can be oxidized with air to regenerate the filter. However, this reaction needs high temperature (typically > 550 °C). In Diesel exhaust, obtaining such a temperature range is a challenge. Hence design of the system needs to be made to periodically increase the DPF temperature to such levels. This can be obtained in two ways. (1) late fuel injection in the cylinder or (2) Direct fuel injection into the exhaust line. In case (1), the exhaust temperature is automatically increased so that

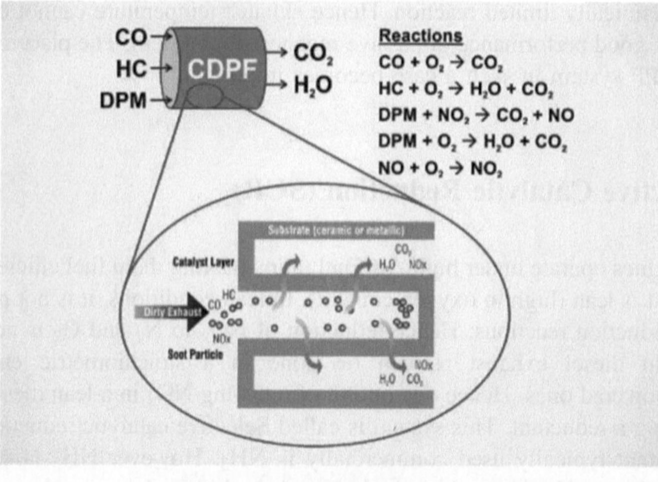

Fig. 4 Schematic of a diesel particulate filter. DPM refers to diesel particulate matter

the DPF experiences required temperature for regeneration. In case (2), the fuel in the exhaust line passes over the DOC. The DOC is designed to burn the fuel resulting in the exotherm. The DOC for such a purpose is expected to be highly thermally stable. Then the resulting exotherm would oxidize the trapped soot in the filter, resulting in regeneration. This form of filter regeneration aided by fuel injections is called 'active regeneration'.

One disadvantage of active regeneration is the fuel penalty it entails. Hence another option for filter regeneration is to oxidize soot with Nitrogen dioxide (NO_2). This is termed 'passive regeneration'. NO_2 is a good oxidizing agent and hence oxidizes soot at lower temperatures (~350–400 °C) than oxygen. Nitrogen oxides coming from engine combustion are typically Nitric oxide NO (90% NO and balance NO_2). Hence in a passive regeneration system, DOC is so designed that NO is oxidized to NO_2 so that NO_2 can further oxidize soot deposited in the Filter, thus regenerating it. It should also be noted that cDPF can be so designed that NO_2 is generated on DPF itself/in addition to DOC (Fig. 4). Due to its relatively low temperature action, the filter can almost continuously be regenerated based on the driving patterns of the vehicle. Having said that, even passive regeneration has a lower temperature limit—both for NO oxidation to NO_2 and soot oxidation with NO_2, which needs at least 200 °C to initiate. If the temperatures are less than these, then the filters get filled up with soot without regeneration. Hence most systems come with an active regeneration backup. This increases the fuel injection intervals hence reduce fuel penalty.

In locomotives, however, it is unlikely that low temperatures would be an issue given the size of the engine and driving patterns. This is particularly true for line-haul locomotives. Hence for line-haul locomotives, another aspect is to be considered. If the temperature of the exhaust is too high, the passive regeneration may still be affected, since the NO_2 generation from NO (on the DOC) is a thermodynamically limited reaction. Hence exhaust temperature cannot over shoot 400 °C for good performance of passive regeneration system. The placement of the DOC + DPF system in such a case becomes important.

8 Selective Catalytic Reduction (SCR)

Diesel engines operate under high Air/Fuel ratios making them fuel efficient. Hence the exhaust is lean (high in oxygen content). In lean conditions, it is not possible to conduct reduction reactions. Hence reduction of NO_x to N_2 and O_2 is not directly possible in diesel exhaust as can be done in a stoichiometric engine like gasoline-powered ones. Hence one option of reducing NO_x in a lean diesel exhaust is by adding a reductant. This system is called Selective catalytic reduction (SCR). The reductant typically used commercially is NH_3. However NH_3 is stored in a precursor form as Urea, stored in a tank. Urea is hydrolyzed on a catalyst to produce NH_3 which then reduces NO_x to N_2 and O_2. The reaction scheme is given below (also see Fig. 5):

$$\text{Urea} \rightarrow NH_3 + CO_2 + H_2O \quad (4)$$

$$NH_3 + NO + O_2 \rightarrow N_2 + H_2O \quad (5)$$

However, it is known that in the presence of NO_2 , the above reaction proceeds much faster and hence in an SCR system, DOC and cDPF are optimized to have $NO:NO_2$ ratio of 1:1.

$$NH_3 + NO + NO_2 \rightarrow N_2 + H_2O \quad (6)$$

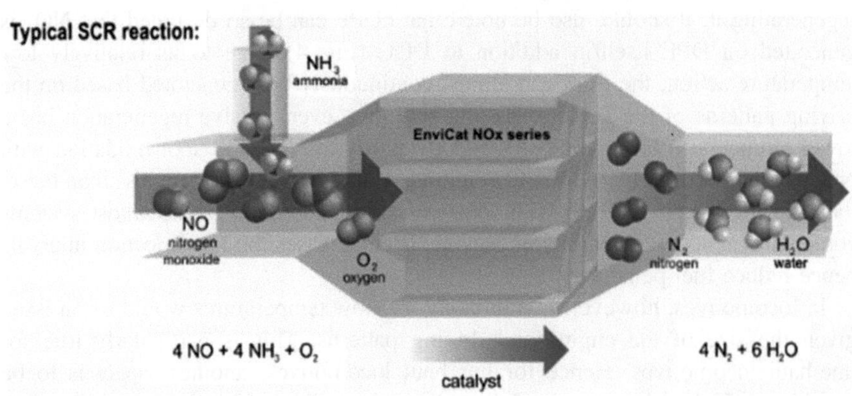

Fig. 5 Schematic of SCR system

A typical SCR consists of Vanadia-Tungsten-Titania (VWT) wash-coat on a ceramic monolith substrate. VWT catalysts have high sulfur tolerance and good low temperature performance. However they do not have high temperature stability.

Hence newer technologies are available with zeolite based materials. Iron doped zeolites have been developed for high temperature activity and Copper doped zeolites for enhanced low temperature activity. New zeolites have made it possible for more broad range performance.

9 Case Studies

While the studies conducted in India for locomotive emission reduction are not readily available, some studies performed overseas may be considered for reference. A recent study was performed by our group company represented by Sud Chemie India Pvt Ltd for catalysts in India. A proprietary DOC was developed for this purpose and the engine was tested. Two separate test cycles were used—one for Line-haul mode and the other for switching mode. The reductions are as shown in Table 9.

Since only DOC was tested in the above system, PM reduction was limited. There was no DPF tested. Obviously, no NO_x data is reported as there is no de-NO_x system used such as SCR.

In another study conducted jointly by Clariant and South West Research institute (SWRI), USA, a 1.5 MW EMD GP20D locomotive was chosen [8]. Here a cDPF was used for the study as well as optimization of fuel injection timing to reduce NO_x. Engine test was done on EPA switch cycle. The results are shown in Table 10 [8].

Table 9 Typical emission reductions obtained on EMD engine using DOC

	Line haul cycle			Switching cycle		
Emissions	HC	CO	PM	HC	CO	PM
% reductions	55	93	30	45	84	55

Table 10 Mass emission results for a EMD GP20D on EPA switch cycle

Test description	HC	CO	NO_x	PM
	g/kW-h	g/kW-h	g/kW-h	g/kW-h
Desired target	0.19	3.22	6.71	0.04
Baseline configuration	1.15	1.63	8.60	0.45
Final configuration	0.01	0.02	6.66	0.01
Percent reduction (%)	−98.7	−99.0	−22.6	−97.9

Here authors note that SCR may be unviable on locomotives due to high fuel consumption on trains. This would made large quantity of urea necessary for locomotives to carry. The feasibility of this has to be looked into before large-scale implementation in IR.

10 Conclusions

Diesel locomotives form majority of the locomotives used in India. However, the emissions of diesel locomotives are not regulated. Like in any other Diesel engine, locomotive engines produce toxic CO, HC, NO_x and PM in their exhaust. First, a suitable norm frame work has to be made so that a detailed study can be undertaken for choosing appropriate after-treatment system.

The authors would like to suggest that Indian Railways may undertake a study for mapping exhaust temperatures and emissions on some pre-selected typical routes for further analysis. This would facilitate a thorough understanding of the type of after-treatment solutions that best suits the Indian locomotive requirements. However, initially DOC only may be a cost effective and practical solution, which can be implemented for immediate purposes, which would result in reduction of >60% HC, >90% CO and >30% PM emissions. Based on the results of the proposed study, the viability of DPF and SCR may be further considered to increase the PM and NO_x reduction from the locomotives of Indian Railways.

References

1. Website https://en.wikipedia.org/wiki/Rail_transport_in_India, October 2016
2. Statistical summary—Indian Railways. http://www.indianrailways.gov.in/railwayboard/uploads/directorate/stat_econ/2014-15/Summary%20Sheet_Eng.pdf
3. Environmental management in India Railways—Stations, Trains and Tracks, 3rd Report, 2014–15, Public Accounts Committee, Ministry of Railways
4. Rail diesel emissions—facts and challenges. International Union of Railways (UIC)
5. Website http://www.uic.org/Emissions#Diesel-emissions, November 2016
6. Website https://www.law.cornell.edu/cfr/text/40/1033.530, November 2016
7. Website https://www.dieselnet.com/standards/cycles/iso8178.php, November 2016
8. Stoos CR, Guliaeff A (2015) Proceedings of the ASME 2015 internal combustion engine division fall technical conference, ICEF 2015, Houston, TX, USA, November 2015

Catalytic Control Options for Diesel Particulate Emissions Including that from Locomotive Engines

Sunit K. Singh, Rohini Khobragade, Govindachetty Saravanan, Avinash K. Agarwal, Ahmed S. AL-Fatesh and Nitin K. Labhasetwar

1 Introduction

Diesel engines have gained an edge over other fuel engines in heavy duty transportation sector owing to their high efficiency, high durability and reliability with low operational costs. Also, in the last few decades, diesel engines have chunked out a growing share in the light-duty vehicle market as well. This rate of growth of diesel engine market also needs careful evaluation of the related environmental impacts due to the emissions of volatile organic compounds (VOCs), nitrogen oxides (NO_x) and mainly that of particulate matter (PM), which is comprised of solid carbon (soot) and unburned carbonaceous compounds [1–3]. PM, in particular, is a major air pollutant with associated health and environmental hazards. The PM emission typically amounts up to 0.2–0.5 mass% of diesel fuel burnt in combustion chambers of engine [4]. However, this could be in a much wider range when considered exceptions related to most recent advanced engines on one hand and those poorly maintained old diesel engines on the other. The fine PMs have been classified into three categories: PM10 (diameter <10 μm), PM2.5 (diameter <0.5 μm) and PM0.1 (diameter <0.1 μm) [5]. Health impacts of these PM include import of toxic metals, like lead, cadmium and zinc, into blood through respiratory system. Also, allergic effects and hypersensitivity due adsorbed chemicals, bacterial and fungal infections

S.K. Singh · R. Khobragade · G. Saravanan · N.K. Labhasetwar (✉)
CSIR-National Environmental Engineering & Research Institute, Nagpur 440020, Maharashtra, India
e-mail: nk_labhsetwar@neeri.res.in

A.K. Agarwal
Engine Research Laboratory, Department of Mechanical Engineering, Indian Institute of Technology Kanpur, Kanpur 208016, India

A.S. AL-Fatesh
Chemical Engineering Department, College of EngineeringKing Saud University, Riyadh 11421, Saudi Arabia

A.K. Agarwal et al. (eds.), *Locomotives and Rail Road Transportation*,
DOI 10.1007/978-981-10-3788-7_9

from live organisms, lung fibrosis and cancer (e.g. asbestos, quartz, chromates and adsorbed organic compounds), irritation of mucous membranes (e.g. acid and alkalis) [6]. PM, also termed as Black Carbon (BC), add to global warming effect by disturbing radiative balance of the atmosphere (of about 1.1 W/m^2), which surmounts to two third of manmade contributors to global climate change other than carbon dioxide, carbon monoxide and methane [7, 8]. Green house gases (GHGs) trap the outgoing infrared radiations and leads to global warming effects. Whereas, BC absorbs both incoming and outgoing radiations, thus contributing more heavily to global warming. Deposition of BC on snow and ice glaciers reduces their reflectivity and increases absorption of radiations leading to accelerated melting. BC similarly modifies properties and distribution of clouds, affecting cloud reflectivity and lifetime ("indirect effects"), stability ("semi-direct effect"), and precipitation [9]. The typical diesel exhaust composition and particulate matter composition is shown in Table 1 [10]. According to a recent study, Indian subcontinent endures higher PM pollution levels that are smaller than 2.5 μm in size (PM2.5) and this exceeds the limit set by WHO [11]. International Agency for Research on Cancer (IARC) has declared PM2.5 as class I cancer-causing agent (carcinogen) in 2013. In summer, these pollution plumes ascend to free atmosphere due to naturally increased buoyancy and efficient ventilation. However, this situation gets adversely aggravated in Gangetic Basin of India, and many other places in the world, during winters as low temperatures help trapping of PMs elevating its surface concentration in the basin. This is evidenced by a strong positive relationship between increased pollution levels and occurrence of dense fog episodes [12].

In the category of heavy-duty transport sector, diesel locomotives have been a significant contributor to total PM emissions in to the environment. Indian railways is the world's second largest railway network with total rail route network of more than 64,000 km. It handles a large fleet of diesel locomotives that are approximately 5000 in number (5714 in number and 52.8% of total locomotives [13]) and growing at the

Table 1 Typical diesel exhaust composition and particulate matter composition [10]

Diesel exhaust composition		Particulate matter composition	
Component	Concentration	Component	Concentration
CO	100–10,000 ppm	Carbon	41%
Hydrocarbons	50–500 ppm, C_1	Unburnt fuel	7%
NOx	30–1000 ppm	Unburnt oil	25%
SOx	Proportional to fuel S content	Sulphate/Water	14%
Particulate matter	*20–200 mg/m^3*	Ash and other	13%
CO_2	2–12 vol.%		
Ammonia	2.0 mg/mile		
Cyanides	1.0 mg/mile		
Benzene	6.0 mg/mile		
Toluene	2.0 mg/mile		
PAH	0.3 mg/mile		
Aldehydes	0.02 mg/mile		

Table 2 Locomotive emission standards set by US-EPA [15]

Tier	Engine manufacture year	HC	CO	NO_x	PM
Line-Haul locomotive emission standards, g/bhp hr					
Tier 0	1973–1992	1.00	5.0	8.0	0.22
Tier 1	1993–2004	0.55	2.2	7.4	0.22
Tier 2	2005–2011	0.30	1.5	5.5	0.10
Tier 3	2012–2014	0.30	1.5	5.5	0.10
Tier 4	2015 or later	0.14	1.5	1.3	0.03
Switch locomotive emission standards, g/bhp hr					
Tier 0	1973–2001	2.10	8.0	11.8	0.26
Tier 1	2002–2004	1.20	2.5	11.0	0.26
Tier 2	2005–2010	0.60	2.4	8.1	0.13
Tier 3	2011–2014	0.60	2.4	5.0	0.10
Tier 4	2015 or later	0.14	2.4	1.3	0.03

rate 250 per year. The diesel fuel consumption in these locomotives surmounts to as high as 2.8 billion litres per year (2,842,739 Kl in 2014–15) [14]. This amount of diesel consumption equates to addition of about half a tonne or more of particulate matter in to the atmosphere. Rail roads operations are categorised under two types of operation: line haul locomotives and switching locomotives [15]. Line haul locomotives refer to cargo transport across long distances. Locomotives for line haul operations are high power engines with 3000–4000 BHP. While switching locomotives are used for assembling and disassembling trains at various locations and engines used are low power engines with 1200–3000 BHP. For locomotive diesel engines, no legislative norms are presently in vogue in India. However, inline with International trends, emissions from Railway locomotives are also expected to be subjected to limits in near future. United States of America (USA) has done the maximum work in the world in the field of measurements, control and legislations for the emissions from Diesel locomotives. United States Environmental Protection Agency (US-EPA) has applied regulations since 1973 in the form of regulatory actions: Tier 0–2 standards and Tier 3–4 standards. Also, in the view of catalytic after-treatment of diesel exhaust emission, US-EPA has regulated the availability of low sulfur diesel fuel for locomotive engines with sulfur limit of 500 ppm effective from June 2007 and sulfur limit of 15 ppm from June 2012. The locomotive emission standards established by US-EPA under these regulatory actions are shown in Table 2[15].

2 Formation, Composition and Structure of Diesel Soot/Particulate Matter

Fuel is introduced as fine droplets in to the combustion chamber. Ultrafine particulate matter, known as soot, is formed due to the incomplete combustion of poorly vaporized large droplets. Incomplete combustion generally occurs in the oxygen deficient region of and around droplets. Local cold spots near the combustion chamber walls with relatively low temperatures also contribute to the incomplete combustion. The precursor molecules like acetylene are formed via incomplete combustion that nucleates and particle growth occurs via reaction with other gaseous components followed by carbonization [16–18]. These diesel particulate matters (PM) are generally composed of carbonaceous soot with condensed polyaromatic hydrocarbons (PAHs). Higher content of aromatic hydrocarbons in diesel fuel enhances the soot emissions due to their facile condensation to PAHs and the aliphatic hydrocarbons support the growth mechanism via consecutive addition of (poly)acetylene [18]. This phenomenon increases with increasing engine speeds due to increased oxygen deficiencies [19].

Diesel PMs or soot, produced by high temperature pyrolysis or combustion of hydrocarbons, often contains a soluble organic fraction (SOF), whose constituents include aromatic compounds as well as various other unburned hydrocarbons [20]. The composition of typical diesel soot is given in Table 3 [21]. It contains at least 1 wt% hydrogen corresponding to empirical composition formula of C_8H for soot [22]. Sulfur is present as adsorbed sulfates, whereas oxygen is present as chemically bonded to form surface functional groups [23] viz. carbonyls, quinines, ethers, phenols, lactones, carboxylic acids and anhydrides. Trace amounts of zinc, phosphorus, calcium, iron, silicon, and chromium are also often detected in emitted particulates. The soot particles are morphologically agglomerates (from few microns to milimeters) of spherules of carbon particles with diameter distribution, typically in the range 10–50 nm [24]. The surface of the spherules has adhering hydrocarbon material or SOF and inorganic material (mostly sulfates). The spherules are called "primary soot particles" and the clusterlike or chain-like soot aggregates are defined as "secondary particles", which are composed of several tens to hundreds of primary spherical particles. Figure 1 shows TEM micrograph of diesel soot showing particles consisting of clusters of spherules.

Micrographs from TEM reveals primary particles to be nearly spherical morphology and laminated surface structure formed by numerous, concentric crystallites [26] (Fig. 2). A hexagonal packing of carbon atoms with face centred arrays, referred to as platelets, has been revealed by X-ray diffractometry. These platelets are arranged in layers to form crystallites with mean spacing of 3.55 nm and typically several hundreds of these crystallites form a primary soot particle. HRTEM image reveals a core-shell structure of primary particles and the platelets model applies to the outer shell. The inner core is composed of carbon network surrounded spherical nucleus indicating its thermodynamic instability due to unorganized structure as compared to outer shell, composed of graphitic crystallites.

Table 3 Elemental composition of diesel soot particulates [21]

Element	C	H	N	O	S[a]
Virgin soot	83.5	1.04	0.24	10.5	1.13
Degassed soot	83.8	0.85	0.22	10.7	0.10

[a]Depends on sulfur content of diesel

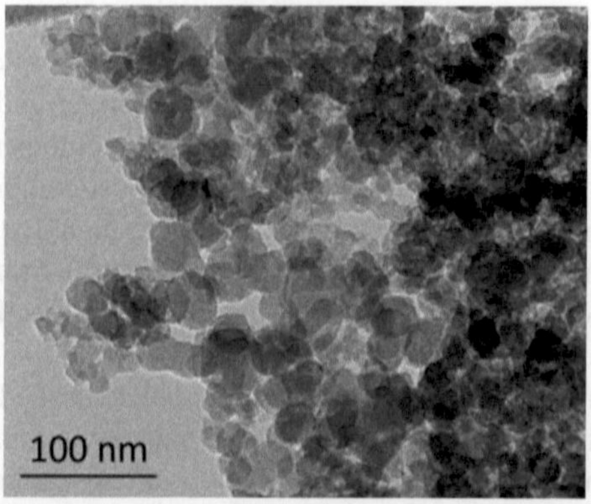

Fig. 1 TEM micrograph of diesel soot showing particles consisting of clusters of spherules (Adapted from Ref. [16])

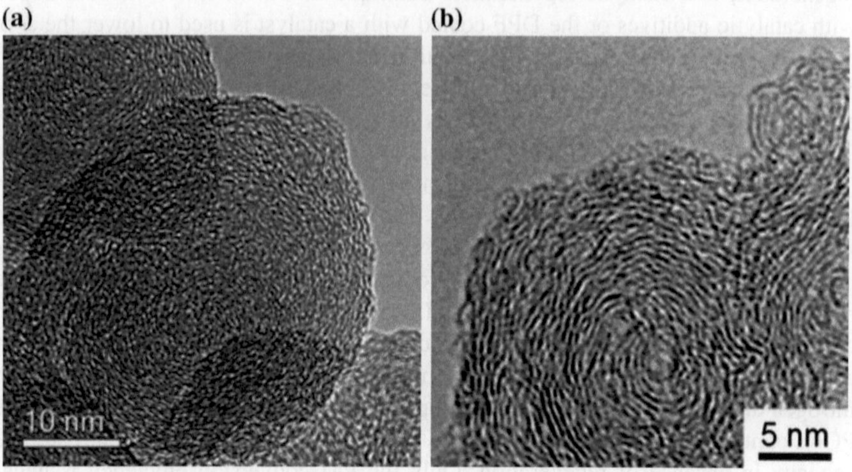

Fig. 2 TEM images of **a** Soot (from Bharat Stage IV diesel engine) showing almost spherical soot primary particles, **b** Soot (from Euro IV diesel engine) consisting of core–shell primary particles with defective bulk and surface structures (Adapted from Ref. [25])

3 Technological Options for Diesel PM Control

Since 80s, significant advancements are observed in the development of emission control technologies for diesel particulates, mainly from automobiles, while very limited efforts have been directed towards locomotives and stationary sources considering their limited contributions to PM emission inventory. Engine designs [27] were given attention for improvements with respect to high pressure fuel injection, small injection nozzle hole area, high swirl ratio, large volume ratio of the piston cavity, improvement of combustion chamber shape and high response turbo-charger. Fuel pre-treatments [28], better tuning of the combustion process [29], use of alternative fuel such as alcohols or biodiesel and modified fuel formulation were attempted as control measures in pre-combustion or during combustion stages. Finally, more promising options in post-combustion stage, use of filtering or non-filtering after-treatment devices were explored. The technological developments with respect to the diesel emission control from automobiles have been briefly discussed here as that can be more or less directly corelated to their application to emission control from locomotives.

Exhaust emissions from diesel engines can be controlled by after-treatment devices that can significantly reduce the tail-pipe emissions. Researchers and engineers are continuously motivated to develop efficient after-treatment systems due to the increasingly strict emission standards. Efficient diesel particulate filtration (DPF) systems with both active and passive regeneration have evolved to remove particulate matter (PM) emissions emitted from diesel vehicles. A passive regeneration is a catalytic regeneration technique in which, either the fuel doped with catalytic additives or the DPF coated with a catalyst is used to lower the soot oxidation temperature. In active regeneration, additional device such as fuel burner, resistive heating coils, or microwave energy is used to provide heat to the diesel exhaust system. However, major efforts are still needed in order to attain high filtration efficiency and optimum regeneration strategies for DPF systems. POCs (Partial Oxidation Catalysts), also known by several names such as Flow through filter, Open particulate filters and Partial flow filters, were designed to collect and store PMs for enough time that allows its catalytic oxidation to harmless gaseous products. POCs also allow free flow-through passage to the exhaust gases while maintaining its PM storage capacity [30]. They are similar to Diesel oxidation catalysts (DOCs) with PM collection and storage capacity like DPFs. The regeneration of these devices is typically accomplished by reactions of soot particles with nitrogen dioxide, that is generated from upstream DOC. However, unlike DPFs, the POCs continue to pass gaseous products even after reaching its maximum storage capacity in absence of regeneration. Only the PM conversion efficiency is negatively affected at this stage. Hence, POCs are counted among the new PM emission control and flow-through filter technologies and have PM control efficiency higher than DOCs but lower than DPFs.

3.1 Diesel Oxidation Catalysts (DOC)

In 1950s exhaust emissions were identified as the main reason for smog. Whereas emissions from indoor forklift trucks, underground mining vehicle, and stationary engines were the major issue in 1960 [31]. In 1970 US Congress passed the Clean Air Act to control the automobile emissions [32]. Catalytic convertor is efficient way to limit the gaseous pollutants. However, high air to fuel ratio in diesel engines limits the use of three-way catalytic convertor, which has been proven as efficient emission control technology for the gasoline engines. To deal with the different complex pollutants emitted from diesel engines and to meet the strict emission norms, multiple control technologies need to develop. Since 1967, DOC has been first time used to control only gaseous emissions [31]. DOC is the first component downstream to the diesel engine to reduce the gaseous pollutants like CO and HC. DOC has the same design as catalytic convertor with multiple parallel channels, which are open from both the end with non-filter design. Monolith or honeycomb is the most commonly used substrate for DOC application due to its several advantages. Monolith structure is available as ceramic and metal with a variable channel, shapes, and size. Monolith is made up of ceramic like Cordierite ($2MgO. 2Al_2O_3. 5SiO_2$), Silicon carbide (SiC).

Nevertheless, Cordierite is very popular and become world standard and used in majority of the vehicles due to the lower cost and better performance. Cordierite ceramic monolith has the large pore size and low surface area with varying channels from 9 to 1200 cpsi for size 13.97 cm radius, and 17.78 cm height of DOC were frequently available. Cordierite with increased cell density from 9 to 1200 cpsi results in increased geometric surface area but at the same time this decreases the diameter of channels, whereas no significant changes observe in pressure drop values of monolith without catalyzed washcoat. The carrier is needed to increase the surface area; however, a carrier is inactive towards the oxidation reaction need to catalyze by impregnating the platinum group metal (PGM). However, catalyzed washcoat results in increased value in pressure drop due to increased thickness of channels. Automobile exhaust usually has high temperature where cordierite shows the little changes in dimension with low thermal expansion coefficient (10 x 10^{-7}/°C), which prevent from cracking due to thermal shock. Temperature changes occur very rapidly, which influence the thermal shock resistance of wash coat. Wash coat expands to large extent at higher temperature compared to cordierite substrate. Cordierite also shows the high mechanical strength, which may resist both axial, and mechanical disturbance occurs in automobiles. The temperature of diesel exhaust for oxidation of soot reaches very high, however, cordierite have the high melting point is over 1300 °C, which is sufficient for environmental applications [33].

Metal monolith is also available as the substrate for environmental applications. Commonly aluminum based alloys like Fe-Cr-Al and Fe-Cr-Al-Ni compositions are used, which show the high thermal resistance. Metallic substrate shows the several advantages over cordierite like high geometric surface area, large open frontal area and high thermal conductivity, which offers the low light-off temperature for cold start, but probably most important is mechanical strength. In a case of metal substrate retro filling is easy simply by welding in vehicles [34]. Although, metallic substrate has some advantages, it also has some durability issue with wash coat and metal support. Metal support can get oxidized at higher temperatures and result in the brittle structure, which some time gets deformed permanently [35].

The catalyst is the key component, which increases the efficiency of DOC and reduces the pollutants. Transition metal like copper, vanadium, nickel was exploited, but their use as DOC was restricted due to the low stability of high temperature and are too sensitive towards the fuel contamination [36]. PGM like platinum (Pt) and palladium (Pd) are most commonly used for DOC application due to their high thermal stability with good performance. Rhodium (Rh) is sometime used to promote NOx oxidation in DOC [37].

The following reactions are most commonly occur in a DOC:

$$CO + 1/2O_2 \rightarrow CO_2 \tag{R1}$$

$$HC + O_2 \rightarrow CO_2 + H_2O \tag{R2}$$

$$\text{Soot (SOF)} \rightarrow CO_2 + H_2O \tag{R3}$$

$$NO_x + O_2 \rightarrow NO_2 \tag{R4}$$

CO oxidized in the presence of excess oxygen and converted into CO_2 whereas HC also oxidized into CO_2 and H_2O, which is shown in reaction R1 and R2. It also oxidized the SOF of soot into CO_2 and H_2O, which vary according to fuel and engine conditions. However, NO_x is not reduced to N_2, but it oxidized into NO_2, which actually promotes the soot oxidation reaction by reducing the temperature. Johnson Matthey showed the DOC as the basic PM control device, with PM reduction with respect to SOF of soot up to 25%. It also reduced the CO and HC up to 70% (Fig. 3) [38].

3.2 *Diesel Particulate Filter (DPF)*

DPF is the 2nd after-control technology, which follows just after the DOC. Since 1999, DPF has been the installed in the vehicle and now a days implemented in large number of cars and heavy duty vehicles [39]. Development in filters and their

Fig. 3 Diesel oxidation catalysts (Adapted from Ref. [37])

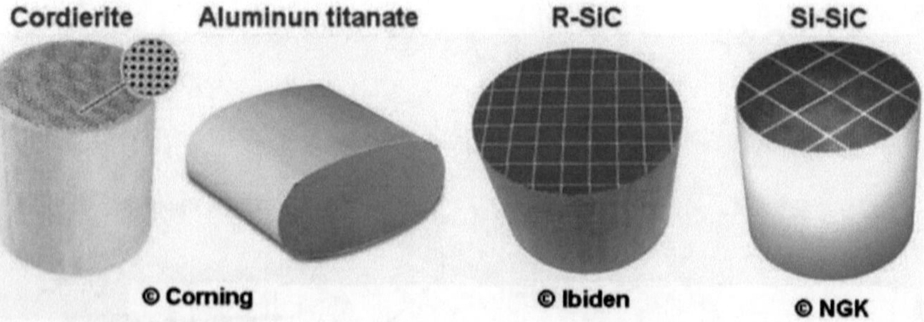

Fig. 4 Examples of commercial ceramic diesel particulate filter (Courtesy to hybrid commercial vehicles Ref. [40])

geometric changes has been observed since 25 years. Most commonly commercially ceramic materials like cordierite (2MgO. $2Al_2O_3$, $5SiO_2$), silicon carbide (SiC), aluminum titanate (Al_2TiO_5) and mullite ($3Al_2O_3$ $2SiO_2$ or $2Al_2O_3$ SiO_2) are used as shown in the Fig. 4. However, metallic filters such as non-woven fiber felt, metal foams and sintered metal powders are also available [40].

DPF are similar to flow through filter as it allows the flow of exhaust gases from the filter. DPF are parallel with alternate channels are plugged on opposite end with the porous wall. The exhaust gases flow through the channels and porous wall allow the gas flow, whereas PM is retained on the wall of the filter. This increases the filter efficiency up to 80% [40, 41]. However, accumulation of soot on the wall of DPF results in the altered filter efficiency and also increased backpressure, which ultimately affects the engine performance. As DPFs are placed after DOC, which is far from the exhaust, where the exhaust gas temperature is not sufficient to regenerate the DPF. To overcome regeneration limitation, catalyst plays important role. The catalyst can decrease the regeneration temperature window from 550–650 °C to 200–350 °C, which is often available with exhaust under the normal operating conditions. Non-catalytic regeneration, which also called active regeneration, is not as efficient as catalytic regeneration also called as passive regeneration. The catalyst is placed on the wall of DPF, where soot particles get trapped and comes in contact with the catalyst and undergo low-temperature oxidation reaction. Therefore, catalyst loaded DPF permits up to 90% filter efficiency with tolerable backpressure. Particulate traps seem to be more realistic regarding economic viability, durability and also can be retrofitted in the vehicles. However, diesel engine exhaust shows the frequent variation regarding temperature of exhaust gas; soot composition etc. and these set the challenges for catalyst application. Still several commercial catalyst based DPF are available, which fulfill the current emission limits. Johnson Matthey trade mark Continuous Regeneration Trap CRT is one of the most popular emission control technologies also available with catalyst

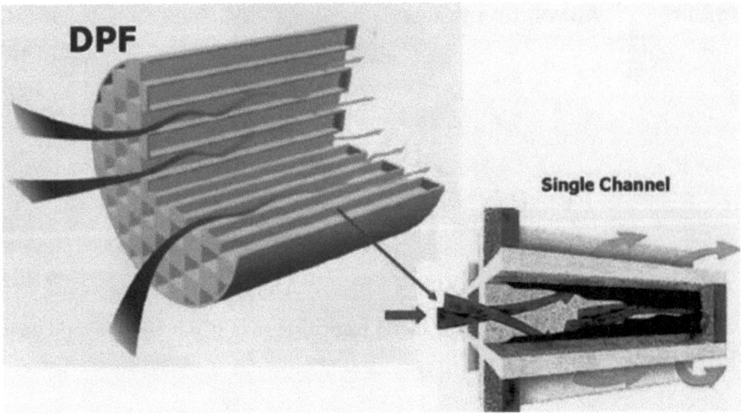

Fig. 5 Diesel particulate wall flow filter (Courtesy to hybrid commercial vehicles Ref. [40])

called as Catalytic Continuous Regeneration Trap (CCRT). Most commonly PGM like Pt and Pd are used as the catalyst in DPF. Pt oxidizes the soot in presence of NO_2 which is formed due to the oxidation of NO_x in DOC. Pt in the presence of NO_2 oxidizes the soot at lower temperature window of 200–300 °C [41–43].

3.3 POCs (Partial Oxidation Catalysts)

POCs are also often known with several names such as Flow through filter, Open particulate filters and Partial flow filters. These devises are designed to collect and store PMs for enough time that allow its catalytic oxidation to harmless gaseous products. POCs also allow free flow-through passage to the exhaust gases while maintaining its PM storage capacity [44] (Fig. 5). They are somewhat similar to DOCs with PM collection and storage capacity like DPFs. The regeneration of these devices is typically accomplished by reactions of soot particles with nitrogen dioxide that is generated from upstream DOC. However, unlike DPFs, the POCs continue to pass gaseous products even after reaching its maximum storage capacity in absence of regeneration. Only the PM conversion efficiency is negatively affected at this stage. Hence, POCs are counted among the new PM emission control and flow-through filter technologies and have PM control efficiency higher than DOCs but lower than DPFs. Traditional POC substrates include ceramic or metallic foams, metal fleece, and wiremesh (Fig. 6) [45].

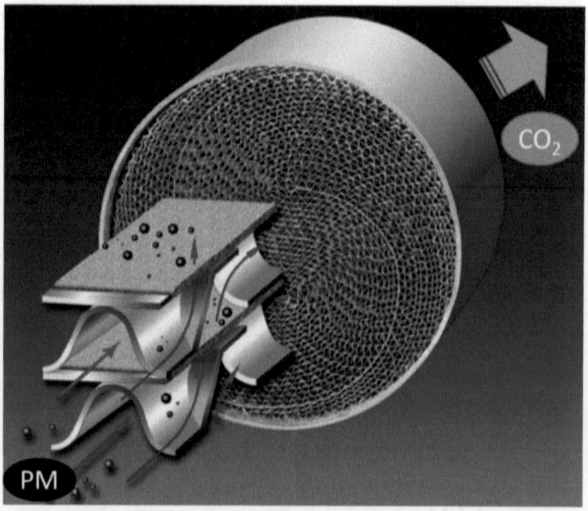

Fig. 6 Metallic flow-through filter made up of corrugated metal foil and layers of porous metal fleece (Adapted from Ref. [46])

4 Diesel Soot Combustion Catalysts

Diesel engines are preferred over gasoline engines due to the use of higher air to fuel (A/F) ratio in internal combustion of diesel as fuel. Higher A/F ratio delivers lower fuel consumption with higher power production and durability associated with diesel engines. However, higher A/F ratio also generates lower combustion temperature resulting in incomplete combustion of fuel and formation of CO, unburnt hydrocarbons and soot. Unburnt hydrocarbons undergo secondary reactions to form polyaromatic hydrocarbons (PAHs) and other volatile organic compounds (VOCs). N_2 in air also leads to formation of higher NO_x emissions from diesel engines. Abatement of NO_x and particulates becomes more difficult due to the lower diesel exhaust temperatures (200–300 °C: light-duty engines and 300–450 °C: heavy-duty engines). These low temperatures demand for the development and use of catalysts in diesel exhaust control system to lower the reaction temperature allowing the effective emission control at available diesel exhaust temperatures. Typically, DOCs are the first component following the engine, where catalytic oxidative control of CO, hydrocarbons, PAHs and VOCs can be achieved with suitable selection of catalysts. Catalysts in DOCs also oxidize NO_x–NO_2 which at second step can be reduced to N_2 using LNT/SCR (lean NO_X trap/selective catalytic reduction) based catalysts. It is observed that reduction of NO_2–N_2 is more feasible than reduction of NO–N_2. DPF and PFF are the third stage of diesel emission control, where diesel PMs are trapped allowing prolonged reaction time availability. Suitable catalysts are also desired in DPF/PFF systems to catalyze low temperature oxidation of PMs and hence effective regeneration of these traps. The catalyst is basically required to operate the following reactions:

Oxidation of CO and hydrocarbons to CO_2 and H_2O:

$$C_xH_y + (1 + y/4)O_2 \rightarrow x\,CO_2 + y/2H_2O \tag{R5}$$

$$CO + 1/2O_2 \rightarrow CO_2 \tag{R6}$$

$$CO + H_2O \rightarrow CO_2 + H_2 \tag{R7}$$

Reduction of NO/NO_2 to N_2:

$$NO(\text{or } NO_2) + CO \rightarrow CO_2 + N_2 \tag{R8}$$

$$NO(\text{or } NO_2) + H_2 \rightarrow CO_2 + H_2O \tag{R9}$$

$$(2 + n/2)NO(\text{or } NO_2) + C_xH_y \rightarrow x\,CO_2 + y/2H_2O + (1 + y/4)N_2 \tag{R10}$$

Diesel PMs emission, being a serious environmental problem, is a subject of stringent emission control regulations. Hence, a continuous research and development activity is progressively focused on exhaust emission after-treatment for developing fuel efficient, lean burn diesel vehicles. This provides continuous stimulus to new ideas in the field of oxidation catalysts and diesel particulate filtration. These efforts require compactness, high volumetric flow rates, high filtration efficiency, low back pressure and acceptable cost of automobile catalysts [47]. The PM oxidation catalyst are aimed for: (1) low cost, (2) high intrinsic activity, to significantly lower particulate oxidation temperature and provide quick particulate combustion and (3) high thermo-chemical stability, to withstand continuous operation over ~100,000 km retaining at least 70% of its original activity.

4.1 Evaluation of Catalytic Soot Oxidation Activity

Temperature-programmed oxidation/combustion experiments (TPO/TPC) are generally used to assess the soot oxidation activity of catalysts. The soot is taken from a diesel engine under different payload conditions and running on different fuel compositions viz. either 100% conventional diesel, ultra low sulphur diesel (ULSD) or blends of ULSD with biodiesel/alcohol. Many researchers also used carbon black as a soot surrogate, which is the best commercially available soot mimicking compound for catalysts screening, since real diesel soot features are heterogeneous. In a typical test, mixture of catalyst and soot/model soot are loaded into a tubular quartz reactor and heated in an oxygen-containing atmosphere at a constant heating rate. The CO_x profiles are recorded in the outlet gases by means of either mass spectrometer or gas chromatograph detectors. However, non-dispersive infrared (NDIR) gas analyzers are more popular option of detection due to the high accuracy for shortest time interval of evaluation in continuous reaction mode. The temperature (T_m) for 100% soot conversion (T_{100}), 50% soot conversion (T_{50}) and the onset temperature (T_{10}) are often used as an index of the catalytic activity. The lower are these temperature, the better is soot oxidation activity of catalyst. Thermo-gravimetric (TG) analysis of catalyst–soot mixtures using mass spectrometer or gas chromatograph detectors also provide useful information such as temperature of the maximum oxidation rate (T_O) and the temperature when soot oxidation begins (T_{onset}) from the mass-temperature plots. In general, the main variables that influence the catalytic performances of soot oxidation catalysts are the soot composition, the soot–catalyst ratio, the type of catalyst–soot contact and the oxygen content (with/without co-oxidant like NO_2, H_2O, etc.) in the feed mixture [48].

4.2 *Soot–Catalyst Interaction Phenomena*

The soot-catalyst interaction/contact is necessary for feasibility of soot oxidation reaction, which is a solid-solid reaction and also solid-gas reaction (Fig. 7). Two types of soot-catalyst contact in laboratory scale studies have been frequently reported in the literature:

(a) *Loose contact*: this condition is simulated by gentle mixing of soot and catalyst with spatula in required ratios and then loading precise amount of this mixture in to the experimental setup as discussed in earlier section. The loose contact mode, as simulated by spatula or in vial mixing of catalyst and soot, was found best simulate the soot and catalyst contact in a DPF environment.
(b) *Tight Contact*: this condition is usually prepared through ball milling (or in a mortar-pestle) to obtain close contact between the catalyst and soot. It maximizes the contact area between soot and catalyst particles providing better combustion efficiency. However, it is less representative of the real contact conditions that occur in a catalytic trap (DPF).

Literature reports have shown that only Cr_2O_3, CuO, Ag_2O, PbO, MoO_3 and Sb_2O_3 show high or moderate activity in tight as well as loose contact mode (Table 4). Other catalysts showing activity in tight contact mode generally fail to show similar activity in loose contact mode.

4.3 *Classification of Soot Oxidation Catalysts*

4.3.1 Pt Based Catalysts

Soot combustion have been reported to be accelerated by many catalytic formulations with noble metals (mainly Pt), alkaline metals and alkaline earth metals, transition metals that can accomplish redox cycles (V, Mn, Co, Cu, Fe, etc.), and internal transition metals. Pt is used in CRT (Continuously Regenerating Trap)

Table 4 Classification of catalyst according to tight and loose contact mode activity (adapted from Ref. [50])

Tight contact		
Hardly any or no activity (T_{comb} > 527 °C) GeO_2, SnO_2, WO_3, Nb_2O_5, MgO, CeO_2, ZnO, BaO, ZrO_2	Moderate activity (527 °C > T_{comb} > 427 °C) Cr_2O_3, NiO, Ag_2O, CuO, CaO, Bi_2O_3, Sb_2O_3, MnO_2, La_2O_3, Fe_2O_3, MoO_3, V_2O_5	High activity (T_{comb} < 427 °C) Co_3O_4, PbO
Loose contact		
Hardly any or no activity (Tcomb > 577 °C) NiO, CaO, Bi_2O_3, MnO_2, La_2O_3, Fe_2O_3, V_2O_3, Co_3O_4	Moderate activity (577 °C > Tcomb > 427 °C) Cr_2O_3, CuO, Ag_2O, PbO, MoO_3, Sb_2O_5	

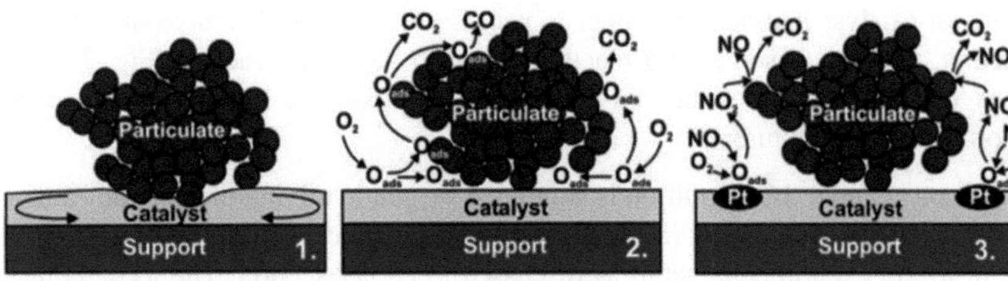

Fig. 7 Diagrammatic representation of soot catalyst interaction via solid-solid and solid-gas reactions (Adapted from Ref. [49])

system as commercial soot removal systems. CRT utilizes a Pt-containing diesel oxidation catalyst in front of DPF, which in addition to controlling HC and CO emissions, also oxidizes NO–NO_2 for low temperature combustion of soot in the downstream filter under the applicable driving conditions. Spill-over of activated oxygen along with NO/NO_2 reactions is assumed as the main mechanism of soot oxidation over Pt based catalysts (Fig. 7(3)). Increase in the surface concentration of oxygen complexes by Pt catalysts followed by reaction with NO_2 enhances the carbon consumption. Pt supported over silica also shows higher activity in synergy to some metals oxides (MoO_3, V_2O_5, K_2O, BaO, etc.) than the single-component catalysts. Various supports like SiO_2, Al_2O_3, ZrO_2, CeO2, $Ce_{0.8}Zr_{0.2}O_2$ (3DOM—three-Dimensionally Ordered Macroporous), were also explored and reported in literature for increased activity of Pt based catalyst.

4.3.2 Ceria-Based Catalysts

CeO_2 alone, or in combination with other metals/metal oxides, was also explored for promising soot oxidation activity. Improvement in soot oxidation activity and stability at high temperatures was observed by substitution of zirconium and many rare earth elements (e.g., La, Pr, Sm, Y, Gd, Tb, Lu, Hf and Nd) in ceria framework. The redox behaviour and the availability of chemisorbed oxygen (α-species) are important factors in the oxidation activity of ceria based materials. Involvement of CeO_2 lattice oxygen (via a Mars–van Krevelen mechanism [51]), in soot oxidation have also been discussed.

4.3.3 Perovskite Catalysts

ABO_3 perovskite-type oxides are reported to catalyze the simultaneous removal of NO_x and diesel soot in the presence of oxygen, and were superior to transition metal simple oxides. Sr-doped $LaMnO_3$ materials were found to be less expensive substitute for Pt in both DOC (Diesel Oxidation Catalysts) and in NSC (NO_x Storage Catalysts). Perovskites (ABO_3) with A sites occupation of La + Sr, La + Li, La + K, or La + Cs and B sites occupation of Co, Mn, Fe, Mn + Fe, Mn + Cu, or Cu + V are the typical combinations reported in wide literature for efficient soot oxidation activity. Higher activity of perovskite based catalysts is assumed to be due to high concentration of suprafacial, weakly chemisorbed oxygen, which actively contributes to soot combustion by spill-over in the temperature range of 300–500 °C. Introduction of alkali metals (Li, K) in to perovskite structure have shown improved activity due to enhancement of the amount of weakly chemisorbed oxygen O^- species [51].

4.3.4 Spinel-Based Catalysts

Ternary AB_2O_4 spinel oxides are also reported for their efficiency as soot combustion catalysts with simultaneous removal of NO_x in the presence of excess oxygen. Catalytic formulations of spinels reported are nanostructured spinel-type oxide catalysts AB_2O_4 (where A = Cu, Co, Mn, Zn and Ba, and B = Cr, Fe and Al). Similar, activity for soot combustion were exhibited by a $CoAl_2O_4$ spinel catalyst and a Pt/Al_2O_3 reference catalyst. The performance of these spinel catalysts was significantly better than that of the mechanical mixtures of these metal oxides. The established catalytic combustion activity of the spinel catalysts is also explained by their higher concentration of suprafacial, weakly chemisorbed oxygen, which contributes actively to soot combustion by spill-over.

4.3.5 Alkaline Metal-Based Catalysts

Alkaline metal were investigated mainly as promoters of other soot oxidation catalysts such as ceria, spinels ($CuFe_2O_4$, $CoFe_2O_4$), perovskites ($LaCoO_3$, $LaFeO_3$, $LaCrO_3$, $LaMnO_3$, $SrTiO_3$), FeO_x–VO_y/Al_2O_3, VO_x/MO (MO = SiO_2, TiO_2, ZrO_2, or Al_2O_3, and Pt/Al_2O_3. Activity of alkaline metals were observed to decrease in an order Cs > Rb > K > Na > Li. In perovskites, such as $LaCoO_3$, alkaline metals (mainly rubidium) substituted at A sites were observed to enhance the activity due to the changes in their basicity, desorption of oxygen, and surface structure.

Promoting effects of potassium were observed on $CuFe_2O_4$-spinel catalyst for soot combustion (optimum loading x = 0.05 in $Cu_{1-x}K_xFe_2O_4$).

The brief details of different soot oxidation catalysts as compiled from exhaustive literature are given in Table 5.

Table 5 Brief compilation of catalysts and their soot oxidation performance reported in literature

Catalysts	$T_{max/50}$ (°C)[a]	Oxidant gas mixture used	Relevant experimental details	Reference
SiO_2 1%Pt/SiO_2 1%Pt/Al_2O_3 1%Pt/ZrO_2	661 312 460 408	10 mol% O_2 + 7 mol% H_2O + 0.1 mol % NO + 0.01 mol % SO_2 in N_2	500 mL min^{-1}; fixed-bed flow reactor; 10 °C min^{-1}; 5 mg of soot + 500 mg of catalyst; carbon black as model soot	Oi-Uchisawa et al. (1998) [52]
γ-Al_2O_3 1%Pt/γ-Al_2O_3 1% Pt/20%Ba/γ-Al_2O_3	650 550 580	0.2 mol% NO_x + 5 mol% O_2 in He	100 mL min^{-1}; fixed-bed flow reactor; 15 °C min^{-1}; 66 mg (m_{soot}: $m_{catalyst}$ = 1:9); Printex U as model soot	Castoldi et al. (2006) [53]
CeO_2 $Ce_{0.76}Zr_{0.24}O_2$ ($Ce_{0.56}Zr_{0.44}O_2$ $Ce_{0.36}Zr_{0.64}O_2$ $Ce_{0.16}Zr_{0.84}O_2$ ZrO_2	518 521 587 596 606 582	0.05 mol% NO_x + 5 mol% O_2 in N_2	500 mL min^{-1}; fixed-bed flow reactor; 10 °C min^{-1}; 100 mg (m_{soot}: $m_{catalyst}$ = 1:4); 30,000 h^{-1}; Printex U as model soot	Atribak et al. (2009) [54]
1%Pt/Al_2O_3 $Ce_{0.6}Zr_{0.4}O_2$ 1%Pt/$Ce_{0.6}Zr_{0.4}O_2$ –	464 460 430 570	1000 ppm NO + 10% O_2 in N_2	500 ml min^{-1}; fixed-bed flow reactor; 10 °C/min; 10 mg soot: 100 mg catalyst; 30,000 h^{-1}; Printex-U as a model soot	Liu et al. (2014) [55]
Mesoporous-Mn_2O_3-Amorphous Mesoporous-Mn_2O_3-bixbyite Mesoporous-ϵ-MnO_2-akhtenskite Mesoporous-MnO_2-Octahedral molecular sieves –	334 323 305 316 493	1000 ppm NO_2 in Ar + Air (1:1)	Soot and catalyst (1:5 wt: wt); WHSV—12 L h^{-1} g^{-1}; fixed-bed reactor; Printex-U as a model soot	Wasalathanthri, et al. (2016) [56]
$Mg_{0.1}La0_{.8}Ce_{0.2}CoO_3$ $MgLa0_{.8}Ce_{0.2}CoO_3$ $Mg_4La0_{.8}Ce_{0.2}CoO_3$ 10 wt% K/ $Mg_{0.1}La0_{.8}Ce_{0.2}CoO_3$ 10 wt% K/ $Mg_4La0_{.8}Ce_{0.2}CoO_3$	432 421 399 385 380	20% O_2 + 80% N_2	50 ml min^{-1}; 10 °C/min; Soot and catalyst (1:9 wt: wt); fixed-bed flow reactor; Printex-U as a model soot	Wang et al. (2016) [57]

(continued)

Table 5 (continued)

Catalysts	$T_{max/50}$ (°C)[a]	Oxidant gas mixture used	Relevant experimental details	Reference
Co_3O_4 $LaCoO_3$ 4.65%Ag/$LaCoO_3$	481 500 437	0.5%NO + 10% O_2 + 89.5%N_2	50 ml min^{-1}; 1 °C/min; soot:catalyst mass ratio 20:80; fixed-bed flow reactor; Printex-U as a model soot	Fan et. al (2016) [58]
$La_{0.6}Ce_{0.4}CoO_3$ $La_{0.6}Ce_{0.4}Ni_{0.5}Co_{0.5}O_3$ $La_{0.6}Ce_{0.4}Cu_{0.5}Co_{0.5}O_3$ $La_{0.6}Ce_{0.4}Fe_{0.5}Co_{0.5}O_3$	436 421 454 433	10% O_2 + 90% N_2	100 ml min^{-1}; 3 °C/min; Soot and catalyst (1:10 wt:wt); fixed-bed flow reactor; Soot from Diesel engine	Liu et.al. (2002) [59]
$SrTiO_3$ $Sr_{0.8}K_{0.2}TiO_3$ $Sr_{0.8}Li_{0.2}TiO_3$ $Sr_{0.8}Cs_{0.2}TiO_3$	470 302 462 303	10% O_2 + 90% N_2	GHSV: 20,000/h; Soot and catalyst (1:5 wt:wt); fixed-bed flow reactor; Printex-U as a model soot	Białobok et al. (2007) [60]
$LaCoO_3$ $La_{0.8}Ce_{0.2}CoO_3$ $La_{0.9}Sr_{0.1}CoO_3$ Co_3O_4 La_2O_3 CeO_2 SrO	485 435 475 455 722 505 >700	10% O_2/He	60 ml min^{-1}; GHSV: 100,000/h; fixed-bed flow reactor; 5 °C/min; soot:catalyst mass ratio 5:95; Printex-U as a model soot	Zhang et al. (2010) [61]
$LaFeO_3$ $La_{0.95}K_{0.05}FeO_3$ $La_{0.9}K_{0.1}FeO_3$ $La_{0.8}K_{0.2}FeO_3$	392 381 373 350	8% O_2 + 92% N_2	50 ml min^{-1}; 5 °C/min; Soot and catalyst (1:9 wt: wt); fixed-bed flow reactor; Printex-U as a model soot	Meng et al. (2011) [62]
La_2CuO_4 Pr_2CuO_4 Nd_2CuO_4 Sm_2CuO_4 Gd_2CuO_4 $La_{1.3}Na_{0.7}CuO_4$	504 507 518 518 520 463	0.2 mol% NO + 5 mol% O_2 in He	50 mL min^{-1}; flow reactor; 2 °C min^{-1}; 108 mg (m(soot): m (catalyst) = 1: 5); Printex U as a model soot	Liu et al. (2008) [63]

4.4 *Soot Oxidation Kinetics*

Considering the energy savings and maintenance costs associated with diesel engines, development of efficient advanced diesel particulate filter systems require to achieve high filtration capacity with high regeneration capabilities. For exploring efficient regeneration strategies, detailed understanding of PM oxidation kinetics and kinetic parameters with respect to different engine operating conditions, at which the concentrations of emission components vary. Considering that soot is composed of ~90 wt% of carbon, its oxidation with O_2 and NO_2 may be assumed to proceed via well known thermodynamic path:

$$C + O_2 \rightarrow CO_2 \quad \Delta H^{\circ}_{298K} = -394 \text{ kJ mol}^{-1} \tag{R11}$$

$$C + 1/2O_2 \rightarrow CO \quad \Delta H^{\circ}_{298K} = -110 \text{ kJ mol}^{-1} \tag{R12}$$

$$C + 2NO_2 \rightarrow CO_2 + 2NO \quad \Delta H = -279 \text{ kJ mol}^{-1} \tag{R13}$$

$$C + NO_2 \rightarrow CO + NO \quad \Delta H = -53 \text{ kJ mol}^{-1} \tag{R14}$$

$$C - H + 2.5\,NO_2 \rightarrow CO + 2.5\,NO + 0.5H_2O \quad \Delta H = -784 \text{ kJ mol}^{-1} \tag{R15}$$

Reaction R11 corresponds to SOFs present on the surface of soot. Multiple parallel and sequential (quasi-) elementary steps have been proposed to illustrate the chemical mechanism of soot oxidation reactions with O_2/NO_2 as discussed by Marsh and Kuo [64], Chen et al. [65], Moulijn et al. [66] and Li et al. [67].

$$C^* + O_2 \rightarrow C(O_2) \text{ or } C(O_2)_m \tag{R16}$$

$$\begin{aligned} &C(O_2)_m \rightarrow C(O) + C(O)_m \text{ or/and } C(O_2)_m \rightarrow C(O)_m + C(O)_m \\ &\text{or/and } C(O_2)_m \rightarrow C(O) + C(O) \end{aligned} \tag{R17}$$

$$C(O) \rightarrow CO \text{ and } C(O)_m \rightarrow CO \tag{R18}$$

$$C(O)_m + C(O)_m \rightarrow C^* + CO_2 \tag{R19}$$

$$C(O)_m + C(O) \rightarrow C^* + CO_2 \tag{R20}$$

$$CO + C(O)_m \rightarrow C^* + CO_2 \tag{R21}$$

$$CO + C(O) \rightarrow C^* + CO_2 \tag{R22}$$

$$O_2 + 2\,C(O) \rightarrow 2\,CO_2 \tag{R23}$$

$$C^* + NO_2 \rightarrow C(ONO) \quad \{C^* \text{ indicates Solid Carbon}\} \tag{R24}$$

$$C(ONO) \rightarrow C(O) + NO \quad \{C(O) \text{ indicates SOCs}\} \tag{R25}$$

$$C(O) + C(ONO) \rightarrow [C(O - ONO)]^{\dagger} \rightarrow CO_2 + NO \tag{R26}$$

$$C(O) \rightarrow C^* + CO \tag{R27}$$

where, free carbon sites are indicated by C*, chemisorbed localised molecular oxygen $C(O_2)$, chemisorbed mobile molecular oxygen $C(O_2)_m$, chemisorbed localised atoms of oxygen C(O) and chemisorbed mobile atoms of oxygen $C(O)_m$. A parametric study for influence of temperature and oxidant concentrations was performed and a mono-dimensional model [68, 69] (Eqs. 1–2) based on global kinetic law was used to generate the intrinsic kinetic parameters of reaction between carbon and O_2/NO_2 based on the reaction [70]:

(1) Without water effect:

$$\frac{1}{\delta m}\frac{d(\delta m)}{dt} = k_{CO_2}X_{O_2} + k_{CO}X_{O_2} + k_{CO_2}X_{NO_2} + k_{CO}X_{NO_2} + (k_{O_2CO_2} + k_{O_2CO})X_{NO_2}^{0.3}X_{O_2} \tag{1}$$

(2) With water effect:

$$\frac{1}{\delta m}\frac{d(\delta m)}{dt} = (k_{CO_2}X_{O_2} + k_{CO}X_{O_2})\left(1 + a\,x_{H_2O}^{\varphi}\right) + k_{CO_2}X_{NO_2}\left(1 + b\,x_{H_2O}^{\mu}\right) + k_{CO}X_{NO_2}\left(1 + c\,x_{H_2O}^{\theta}\right) + (k_{O_2CO_2} + k_{O_2CO})X_{NO_2}^{0.3}X_{O_2} \tag{2}$$

where X_i are molar fractions, while k_j are intrinsic kinetic constants expressed as s^{-1} which are intended as grams of soot oxidized per initial gram of soot and per unit time. The activation energy of the O_2 mediated oxidations leading to CO_2 is 127 kJ/mol, while the one that leads to CO is 170 kJ/mol. In a similar way to the NO_2 mediated mechanism, the activation energy is 39 and 66 kJ/mol for CO_2 and CO, respectively. The full set of pre-exponential values and constants in Eqs. (1–2) can be found in Zouaoui et al. [70].

5 Commercial Technologies

5.1 Flow-Through Diesel Oxidation Catalysts

Flow-through diesel oxidation catalysts were the first commercial devices of diesel particulate emission control, which became mandatory in diesel engines in 1996 in US and 1998 in Europe. These convertors were similar to conventional catalytic convertors of gasoline engines with some modifications in catalyst composition to optimize the catalyst activity under lean conditions. These did not work as a trap for soot particles but CO and hydrocarbons were efficiently oxidised with $\sim 5\%$ particulate abatement at gas hourly space velocities of 50,000–300,000 h^{-1}.

5.2 PSA (Peugeot-Citröen Societé d'Automobiles) System

This system was based on particulate filtration and trap methodology using SiC wall-flow monolith to achieve high filtration efficiency and superior high-temperature and thermal-shock resistance. In this system, active regeneration of filter is done by fuel post-injection, enabled by multi-jet common-rail engines, to increase exhaust gas temperature for combustion of diesel PMs. This system also used CeO_2 based fuel borne catalytic system where Ce-fuel additive leads to formation of CeO_2 particles well embedded in the structure of diesel particulate and thus in very good contact with the soot, which lowers ignition temperatures.

5.3 Continuously-Regenerating-Trap (CRT) System

The technology of continuous regeneration of monolithic wall-flow filters by NO_2 generated in an upstream oxidation catalyst was developed by Cooper and Thoss [71], and it has been commercialized as the "Continuously regenerating trap (CRT)" system [72]. The upstream oxidation catalyst converts about 90% of the HC and CO present in the exhaust gas and promotes the NO_x–NO_2 conversion. This trap has the ability to continuously regenerate, provided its operating temperature is kept in the range 200–450 °C. With efficiently working upstream oxidation catalyst and avoiding extreme temperature gradients within the trap, a satisfactory performance of CRT over 600,000 km has been reported [73].

Catalytic DPF system also called as Catalytic continuously regenerating trap (CRT which is trademark of Johnson Matthey) system is used with additional coating of catalyst in the filter pores. Catalyst loaded on DPF helps to reoxidise the reduced NO, which increases the NO_X/PM ratio and lower down the soot oxidation temperature up to 200 °C, thus enhancing the soot oxidation activity [74].

5.4 Toyota Motors System

In this system, diesel soot filtration device (DPF) is deposited with a layer of an "active oxygen" storage alkali metal oxide hosting Pt. Gas phase conversion of NO_x to surface nitrate species results in formation of active oxygen. Decomposition of these nitrates at the interface of soot and active layer produces highly reactive adsorbed oxygen atoms oxidizing deposited soot at temperatures ≥ 300 °C. The layer of alkali metal oxide simultaneously acts as NO_x trap and particulate abatement system [75].

6 Post-Combustion PM Control Options for Locomotives

As discussed, diesel provides significant fuel economy and durability for light-duty and heavy-duty transportation vehicles as well those stationary engines. However, they are associated with significantly large emissions of particulate matter (PM) and oxides of nitrogen (NO_x), and lesser amounts of hydrocarbon (HC), carbon monoxide (CO), and toxic air pollutants. Locomotives in many countries including India are significant contributors to air pollution, although their contribution has been drastically reduced with massive electrification of locomotives during the last two decades or so. Although, exhaust emission norms are presently not in force for locomotives in India, these are expected in near future. It will be however, not unfair to mention that the experience of other countries can be very much used to design these emission norms, as experienced in case of those for automobiles (mostly based on European emission standards). Similarly the locomotive emission control technologies in vogue will provide excellent knowledgebase for developing such options for Indian locomotives. Moreover, the above discussed PM control options for automobiles would need design modifications for their applications in locomotives. These technologies involve diesel oxidation catalysts (DOCs) and diesel particulate filters (DPFs) for controlling diesel PM emissions, and lean NO_x catalysts and selective catalytic reduction (SCR) catalysts for reducing NO_x emissions. However, these technologies may have challenges for their applications in locomotives due to operational differences in on-road diesel engines and railways locomotives. Line-Haul locomotives are designed to operate at low speeds without transients as on on-road diesel engines. However, these technologies are under active development and feasibility studies on locomotives in US and other developed nations are well in progress. In this way, it will be advisable to immediately explore the diesel emission control from Indian locomotives considering the strong background available on emission control from diesel cars and heavy duty vehicles, as well as commercial technologies available fro locomotives, which can be adopted with required modifications for a significant gain in PM emission control.

Acknowledgements This book chapter is written based on the R&D work being carried out under the International Scientific Partnership Programme (through ISPP#0057) of King Saud University, Riyadh, Saudi Arabia for establishing research collaboration between King Saud University and CSIR-NEERI. One of the authors (RK) would like to thank UGC—JRF for the fellowship.

References

1. Aneggi E, de Leitenburg C, Trovarelli A (2013) In: Trovarelli A, Fornasiero P (eds) Catalysis by Ceria and related materials, 2nd edn. Imperial College Press, London, pp 565–621
2. Heck RM, Farrauto RJ, Gulati ST (2006) Catalytic air pollution control: commercial technology, 3rd edn. Wiley-VCH, Hoboken, New Jersey, p 518

3. Lox ESJ (2008) In: Ertl G, Knözinger H, Schüth F, Weitkamp J (eds) Handbook of heterogeneous catalysis, 2nd edn. Wiley-VCH, Weinheim, pp 2274–2344
4. Heywood JB (1988) Internal combustion engine fundamentals. McGrawHill International, New York
5. Xi J, Zhong BJ (2006) Chem Eng Technol 29:665
6. Niessner R (2014) Angew Chem Int Ed 53:12366
7. Pueschel RF, Kinne SA (1995) Sci Total Environ 161:811
8. Bond TC, Doherty SJ, Fahey DW, Forster PM et al (2013) J Geophys Res—Atmos 118:5380
9. Report to Congress on Black Carbon (2012) US Environmental Protection Agency, EPA-450/R-12-001, March 2012
10. Prasad R, Bella VR (2010) Bull Chem React Eng Catal 5:69
11. Shamjad PM, Tripathi SN, Pathak R, Hallquist M, Arola A, Bergin MH (2015) Environ Sci Technol 49:10474
12. Chakraborty A, Bhattu D, Gupta T, Tripathi SN, Canagaratna MR(2015) J Geophys Res—Atmo 120, 9006
13. Facts & Figure-2014–15, Indian Railways Statistical Publications 2014–15, Ministry of Indian Railways (Railway Board)
14. IR Annual Statistical Statements 2014–15, Indian Railways Statistical Publications 2014–15, Ministry of Indian Railways (Railway Board)
15. Taylor E, McMillan A (2014) Air quality management: Canadian perspectives on a global issue, Springer Science, p 229
16. Bueno-López A (2014) Appl Catal B 146:1–11
17. Frenklach M (2002) Phys Chem Chem Phys 4:2028
18. Xi J, Zhong BJ (2006) Chem Eng Technol 29:665
19. Taghavifar H, Khalilarya S (2015) S Jafarmadar Energy Sci Eng 3:360
20. Shigeki D et al (2000) JSAE Rev 21:303
21. Marcucilli F, Gilot P, Stanmore BR, Prado G (1994) 25th Symposium (International) on Combustion, The Combustion Institute, Philadelphia
22. Palmer HB, Cullis HF (1965) Chem Phys Carbon 1:265
23. Setiabudi A, Makkee M, Moulijn JA (2004) Appl Catal B 50:185
24. Stanmore BR, Brilhac JF, Gilot P (2001) Carbon 39:2247
25. Su DS, Serafino A, Miller J-O, Jentoft RE, Schlçgl R, Fiorito S (2008) Environ Sci Technol 42:1761
26. Haynes BS, Wagner HG (1981) Prog Energ Combust 7:229
27. Johnson TV (2002). Diesel emission control: 2001 in review, SAE Paper, vol. 2002-01-0285
28. Kittelson DB (1998) Engines and nanoparticles: a review. J Aerosol Sci 29:575
29. Mayer A, Matter U, Scheidegger W, Czerwinsky J, Wieser M, Kieser D, Weidhofer J (1998) Diesel nano-particulate emissions: properties and reduction strategies, SAE Paper, vol. 980539
30. Vakkilainen A, Lylykangas R (2004) Particle oxidation catalyst (POC) for diesel vehicles, SAE Technical Paper, 2004-28-0047. doi:10.4271/2004-28-0047
31. Heck RM, Farrauto RJ (1995) Catalytic air pollution control. Van Nostrand Reinhold, New York, NY
32. United States Environmental Protection Agency. Clean Air Act. http://www.gpo.gov/fdsys/pkg/USCODE-2008-title42/pdf/USCODE-2008-title42-chap85.pdf. Accessed 23 August 2010
33. Heck RM, Farrauto RJ, Gulati ST (2009) The preparation of catalytic materials: carriers, active component, and Monolith Substarte. In: Catalytic air pollution control commercial technology, 3rd edn, Chapter 2 John Wiley & Sons
34. Maattanen M, Lylykangas R (1990) Mechanical strength of metallic catalytic convertor made up of precoated foil SAE 900505
35. Gulati S, Summers J, Linden D, Mitchell, K Impact of washcoatformulation on properties and performance of cordierite ceramic convertor" SAE912370 (1991d)

36. Yu-Yao Y-F, Kummer JT (1977) J Catal 46:388–401
37. Twigg M (2006) Catal Today 117:407–418
38. Matthey J Stationary emission control www.jmsec.com
39. Johnson TV (2008) Diesel emission control in review SAE Tech Paper No 2008-01-0069
40. Konstandopoulos A.G, et al (2000) SAE Tech Paper No 2000-01-1016 March 2000; Konstandopoulos A.G., Report- Technology Evaluation Report for Diesel Engine After-treatment System, Hybrid Commercial Vehicles, June2013.
41. Ambrogio M, Saracco G, Specchia V (2001) Chem Eng Sci 56:1613–1621
42. Website of Corning Environmental Technologies Company (2003) www.corning.com/ enviromentaltechnologies
43. Website of Johnson and Matthey Company, Environmental Catalysis Technologies Divison, http://www.jmdpf.com/diesel-exhaust-filter-system-CRT-NRMM-johnson-matthey
44. Vakkilainen A, Lylykangas R (2004) SAE Technical Paper 2004-28-0047
45. http://www.meca.org/technology/technology-details?id=6&name=Particulate%20Filters
46. Emission Control Technologies for Diesel-Powered Vehicles—Report (2007), Manufacturers of Emission Controls Association
47. Fino D, Specchia V (2004) Chem Eng Sci 59:4825
48. Uchisawa J, Obuchi A, Nanba T (2014) In: Duprez D, Cavani F (eds) Handbook of advanced methods and processes in oxidation catalysis. Press, Imperial College, pp 25–50
49. Fino D, Specchia V (2008) Powder Technol 180:64–73
50. Frennet A, Bastin J.-M (1995) Catalysis and automotive pollution control III, Stud Surf Sci Catal, Elsevier Science, pp 555
51. Lox ESJ (2008) In: Ertl G, Knözinger H, Schüth F, Weitkamp J (eds) Handbook of heterogeneous catalysis, 2nd edn. Wiley-VCH, Weinheim, pp 2274–2344
52. Oi Uchisawa J, Obuchi A, Zhao Z, Kushiyama S (1998) Appl Catal B 18:L183–L187
53. Castoldi L, Matarrese R, Lietti L, Forzatti P (2006) Appl Catal B 64:25–34
54. Atribak I, Azambre B, Bueno López A, García-García A (2009) Appl Catal B 92:126–137
55. Liu S, Wu X, Lin Y, Li M, Weng D (2014) Chin J Catal 35:407–415
56. Wasalathanthri ND, SantaMaria TM, Kriz DA, Dissanayake SL, Kuo CH, Biswas S, Suib SL (2017) Appl Catal B 201:543–551
57. Wang Lei, Fang Shuqing, Feng Nengjie, Wan Hui, Guan Guofeng (2016) Chem Eng J 293:68–74
58. Fan Q, Zhang S, Sun L, Dong X, Zhang L, Shan W, Zhu Z (2016) Chin J Catal 37:428–435
59. Liu Z, Hao Z, Zhang H, Zhuang Y (2002) J Chem Technol Biotechnol 77:800–804
60. Białobok B, Trawczyński J, Rzadki T, Miśta W, Zawadzki M (2007) Catal Today 119:278–285
61. Zhang R, Luo N, Chen B, Kaliaguine S (2010) Energy Fuels 24:3719–3726
62. Meng X, He F, Shen X, Xiang J, Wang P (2011) Ind Eng Chem Res 50:11037–11042
63. Liu J, Zhao Z, Xu C, Duan A (2008) Appl Catal B 78:61–72
64. Marsh H, Kuo K (1989) Kinetics and catalysis of carbon gasification. In: Marsh H (ed) Introduction to carbon science, Butterworths, London p. 107
65. Chen SG, Yang RT, Katpeijn F, Moulijn JA (1993) Ind Eng Chem Res 12:2835
66. Moulijn JA, Kapteijn F (1995) Carbon 33:1155
67. Li C, Brown TC (2001) Carbon 39:725
68. Jeguirim M, Tschamber V, Brilhac JF, Ehrburger P (2005) Fuel 84:1949
69. Schejbal M, Stepanek J, Marek M, Kocı P, Kubıcek M (2010) Fuel 89:2365
70. Zouaoui N, Labaki M, Jeguirim M (2014) Comptes Rendus Chimie, 17, 672–680
71. Cooper BJ, Thoss JE (1989) The role of NO in diesel particulate emission control. SAE paper 890404
72. Cooper BJ, Radnor HJ, Jung W, Thoss JE (1990) Treatment of diesel exhaust gases. US patent 4,902,487, 1990
73. Allanson R, Cooper BJ, Thoss JE, Uusimäki A, Walker AP, Warren JP (2000) European experience of high mileage durability of continuously regenerating diesel particulate filter technology, SAE Paper, vol. 2000-01-0480

74. Website of Johnson and Matthey Company, Environmental catalysis technologies divison. http://www.jmcsd.com/CCRT_0500501__CARB_verified__new_temp.pdf
75. Nakatani K, Hirota S, Takeshima S, Itoh K, Tanaka T (2002) Simultaneous PM and NOx reduction system for diesel engines, SAE Paper, vol. 2002-01-0957

Soot Formation in Turbulent Diffusion Flames: Effect of Differential Diffusion

Rohit Saini, Manedhar Reddy and Ashoke De

Abstract The present study aims at understanding the influence of differential diffusion on the evolution of flow field and soot volume fraction. For this purpose, two different turbulent diffusion flames (Delft flame III, pilot stabilized natural gas flame and an unconfined turbulent lifted ethylene/air jet flame) are investigated for flow field and soot predictions. The presumed shape multi-environment Eulerian PDF (EPDF) is used as turbulence-chemistry interaction model, while the radiative heat-transfer equation is modeled based on the truncated series expansion in spherical harmonics (P1 approximation). Two approaches are used to model diffusivity, a unity Lewis Number, and multi-component diffusion approach. For predicting soot evolution, an acetylene-based semi-empirical model (Moss-Brookes model and Method of Moments (MOM) model is used. Both the models consider the inception, surface growth and oxidation processes of soot. The influence of temperature is included in terms of the effective absorption coefficient and turbulence-chemistry interaction effects are included in terms of a single variable PDF in terms of temperature. The predictions elucidate the influence of temperature on soot volume fraction. Differential diffusion results in an increase in the soot volume fraction.

Keywords Differential diffusion · Soot perditions · Diffusion flames · Transported PDF

Nomenclature

f	Mixture fraction
Ω_D	Diffusion collision integral
u	Axial velocity
ρ	Mixture density
J_i	Molecular diffusion flux
ψ	Composition space vector

R. Saini · M. Reddy · A. De (✉)
Department of Aerospace Engineering, Indian Institute of Technology Kanpur, Kanpur 208016, India
e-mail: ashoke@iitk.ac.in

A.K. Agarwal et al. (eds.), *Locomotives and Rail Road Transportation*,
DOI 10.1007/978-981-10-3788-7_10

S_k	Reaction rate for species 'k'
D_{ij}	Diffusion coefficient
P_{abs}	Absolute pressure
T	Temperature
G_λ	Incident radiation
λ	Wavelength
n	Moment order

Abbreviations

PAH	Poly-cyclic aromatic hydrocarbons
PDF	Probability density function
EMST	Euclidean minimum spanning tree
EDC	Eddy dissipation concept
IEM	Interaction by exchange with mean
DNS	Direct numerical simulation
LES	Large eddy simulation
SIMPLE	Semi-implicit method for pressure-linked equations
RANS	Reynolds-Averaged-Navier-Stokes

1 Introduction

Soot is the condensed carbon particles resulting from the incomplete combustion of hydrocarbons, indicates poor utilization of fuel. Soot acts as a strong source of radiation and accurate prediction of the soot formation is important in better design of internal combustion and gas turbine engines [1]. One of the major concerns with the theoretical prediction of soot formation is the complex physics, which makes systematic understanding difficult and makes modeling a challenging task. Soot formation and oxidation phenomenology mechanisms have undergone several strides towards making accurate quantitative estimates in combustion systems. The evolution process of soot has been reviewed by Haynes and Wagner [2], Bockhorn [3], Kennedy [4] and have been classified into four major sub-processes: such as the formation and conglomeration of PAH (particle inception or nucleation), the surface reaction of particles and oxidation of particles. Combustion models are required to resolve stiff reactions to accurately predict the formation of PAH, considered as the building block of soot. The temporal fluctuations in turbulent combustion are generally resolved using a scalar PDF approach. Soot formation is relatively slow to the chemistry of flame and soot evolution occurs over a wide range of scales in turbulent flames, imposing additional constraints on the combustion models. Soot is strongly influenced by differential diffusion and the interaction of soot, radiation, and turbulence, further adding complication to the system.

The often assumption of equal molecular diffusivities of the participating species is assumed while modeling gaseous diffusion flames. Furthermore, Lewis number is employed as unity, i.e. the thermal diffusivity is equivalent to the mass diffusivity. These assumptions are provided to attain a single conserved scalar, the mixture fraction, which uniquely determines the species mass fractions and the enthalpy if the Damkohler numbers are in sufficiently large regimes [5]. Whereas in most of the practical static and dynamic propulsion devices operate under the non-premixed combustion, and the respective molecular diffusivity of the participating species varies predominantly. This phenomenon makes it necessary to account the molecular transport coefficients which differ from each other quite significantly, leading to provoke the inclusion of the effect of the different diffusion. Mardani et al. [6] used EDC as turbulence-chemistry interaction model and studied the augmenting sensitivity of the molecular diffusion in the MILD combustion regime under different oxygen concentration and jet Reynolds numbers. DNS data of the turbulent premixed methane-air combustion has been analyzed using PDF mixing models in conjunction with differential diffusion [7]. Further, Richardson and Chen [7] suggested the potential of IEM mixing model in describing the effects of differential diffusion at a lower computational cost as compared to the EMST model. De and Dongre [8] assessed different turbulence-chemistry interaction models including MEPDF method using IEM closure in MILD combustion regime and obtained satisfactory predictions of flame statistics as compared to EDC model. Martinez and Rigopoulos [9] used LES-CMC tool to model the effect of differential diffusion and computing the evolution of soot formation in atmospheric methane flames. Pitsch et al. [10] included differential diffusion to an unsteady Lagrangian flamelet model in determining the evolution of soot, and the unity Lewis number calculations over predicted soot on the fuel rich side. Kronenburg and Bilger [11] accounted differential diffusion in their CMC calculations. The model tended to move species faster than the unity Lewis number and required solving of addition equations for each differentially diffusing scalar. Recently, Wang [12] explored linear and non-linear molecular differential diffusion models in the laboratory-scale turbulent non-premixed jet flames with varying Reynolds numbers.

Modeling of soot formation in turbulent diffusion flames have employed from RANS to LES to DNS. Empirical or semi-empirical soot models were widely used in most of the studies. LES study on soot formation in Delft Flame III was carried out by Mueller et al. [13] and Donde et al. [14], predictions were in reasonable agreement with the experimental measurements. The uncertainty in the PAH mechanism of ethylene resulted in the early evolution of soot in the burner. The study also explained the effect of sub-filter variance on soot formation. A sectional approach was used by Kohler et al. [15] to predict the soot formation in an unconfined lifted turbulent ethylene/air jet flame, and the predictions indicated the early evolution of soot compared to experimental measurements. The study used soot particle size distribution to explain soot growth, coagulation and oxidation along the center line. Lignell et al. [16] included the soot-turbulence interactions. They have observed independence in flame surface and soot evolution; soot motion away from or towards the flame had equal probability. Mueller and Pitsch [13] used Hybrid Method of Moment (HMOM) method to address the intermittencies in soot-turbulence interactions using presumed soot subfilter PDF model. The present study is aimed at understanding the effects of unity Lewis number and differential molecular diffusion on flow field and soot evolution using semi-empirical and Method of Moment model. The prominent advantage of the method of moment model over semi-empirical soot model is its ability to parameterize the brief insights of soot particle size distributions leading to the motivation of the current study. Comparison with experimental measurements can be used to understand the interactions between mixture fraction, temperature and soot particles distribution. The study also examines the effect of soot-turbulence-radiation interactions.

2 Numerical Methods

In the present work, presumed shape multi-environment Eulerian PDF (MEPDF) is used to model the turbulence-chemistry interactions. The chemical equation is represented by multiple reactive scalars. Finite numbers of Delta functions are used to describe the shape of joint composition PDF, and the equation for a single point probability in a turbulent reacting flow is given as:

$$\begin{aligned} &\frac{\partial \rho f_\varphi}{\partial t} + \frac{\partial}{\partial x_i}\left[\rho u_i f_\varphi\right] + \frac{\partial}{\partial \psi_k}\left[\rho S_k f_\varphi\right] \\ &= -\frac{\partial}{\partial x_i}\left[\rho\left\langle u_i^{''}|\psi\right\rangle f_\varphi\right] + \frac{\partial}{\partial \psi_k}\left[\rho\left\langle\frac{1}{\rho}\frac{\partial J_{i,k}}{\partial x_i}|\psi\right\rangle f_\varphi\right] - \frac{\partial}{\partial \psi_k}\left[\rho S_k f_\varphi\right] \end{aligned} \tag{1}$$

where, f_φ is the joint probability density function (PDF) of species composition and enthalpy for a single point. In RHS of Eq. (1), the first term is turbulent scalar flux and is closed by a turbulence model. The second term is the micro-mixing, where IEM model is used for the closure and the third term is the reaction source term. The multicomponent diffusion is modeled through the kinetic theory, assuming zero thermal diffusion and is given as [17] Eq. (2), whereas the diffusion flux is represented using Eq. (3):

$$D_{ij} = 0.00188\frac{\left[T^3\left(\frac{1}{M_{w,i}} + \frac{1}{M_{w,j}}\right)\right]^{1/2}}{P_{abs}\sigma_{ij}^2\Omega_D} \tag{2}$$

$$J_i = -\rho D_{i,m}\nabla Y_i \tag{3}$$

2.1 Radiation Modelling

The radiative transfer equation is assumed as a truncated series expansion in spherical harmonics (*P1* approximation [18]) based on optically thick approximation, defined using the following equation:

$$-\nabla \cdot \left(\frac{1}{3a_\lambda} \nabla G \right) = a_\lambda (4\pi i_\lambda - G_\lambda) \tag{4}$$

In Eq. (4), the left-hand side is the gradient of radiative heat flux and the right-hand side is the source of the radiative heat added to the energy equation. The quantity 'a_λ' is the effective absorption coefficient accounting for mixture of soot and an absorbing (radiating) gas. WSGG model with four fictions gases is considered to represent the non-gray medium. The further details of the approach can be found in Yadav et al. [19].

2.2 Soot Modelling

In the current study, a semi-empirical and Method of Moment (MOM) models are used extensively to study the soot formation phenomenon in the 'Delft Flame III' and 'Lifted turbulent ethylene/air flame'. The governing equations and various sub-processes of soot formation used in the current study are discussed in the subsequent section.

2.2.1 Semi-empirical Approach

In the Moss-Brookes model [20], the evolution of soot is described by the transport equations in terms of soot volume fraction and nuclei concentration as:

$$\frac{\partial \rho Y_{soot}}{\partial t} + \nabla \cdot (\rho \bar{v}) Y_{soot} = \nabla \cdot \left(\frac{\mu_t}{\sigma_{soot}} \nabla Y_{soot} \right) + \frac{dM}{dt} \tag{5}$$

$$\frac{\partial \rho b^*_{nuc}}{\partial t} + \nabla \cdot (\rho \bar{v}) b^*_{nuc} = \nabla \cdot \left(\frac{\mu_t}{\sigma_{nuc}} \nabla b^*_{nuc} \right) + \frac{1}{N_{norm}} \frac{dN}{dt} \tag{6}$$

The soot formation is defined in terms of soot particle number density (*N*) and mass density (*M*). The net gas phase particles nucleation and coagulation give the source term of instantaneous number density. The net nucleation, growth, and oxidation of soot particles gives the source term for the instantaneous soot mass concentration. Currently, partial equilibrium approach is used for calculation of OH and O_2 molecules for estimating the soot oxidation. Further, details of the modeling approach can be found in Reddy et al. [21].

2.2.2 Method of Moments (MOM) Approach

Under this approach, the transport equations are used to get the soot mass fractions by means of the method of moment approach with interpolated closure [22]. In the present study, only first three moments have been considered for the evolution of soot statistics and distribution and closure is achieved through logarithmic interpolation. The concentration moment of the particle number density is formulated for given number of moments and is characterized as:

$$M_n = \sum_{i=1}^{\infty} m_i^n N_i \tag{7}$$

where M_n and N_i represent the n-th moment of soot size distribution and particle density of the size class 'i' respectively. Accordingly, $n = 0$ and 1, corresponds to the total particle density and total mass of the particles. In order to compute soot mass concentrations, the following transport equation for a given number of moments is solved in the physical space:

$$\frac{\partial(\rho M_n)}{\partial t} + \nabla \cdot (\rho \vec{v} M_n) = \nabla \cdot \left(\frac{\mu_{eff}}{\sigma_t} \nabla M_n \right) + S_n \tag{8}$$

where, the source term 'S_n' on the RHS includes statistics related to the evolution of sub-processes, i.e. nucleation, coagulation, surface growth and oxidation, respectively, and can be written as follows:

$$\frac{dM_n}{dt} = S^{nuc.} + S^{coa.} + S^{surf.Growth + Oxid.} \tag{9}$$

The closures of these source terms are discussed in detail as follows:

Nucleation

Nucleation can be modeled as a coagulation procedure between two soot precursor species. In the present study, Acetylene is included as the primary soot precursor; and this precursor (Acetylene) further acts as a building block in the formation of the PAH molecules [23].

The nucleation source term for the nth moment is computed as follows:

$$S_n^{nuc.} = \gamma C_{nuc.} \sqrt{T} [X_p]^2 \tag{10}$$

In Eq. (10), $[X_p]$ is the molar concentration of the precursor species and $C_{nuc.}$ is the constant. γ is the sticking coefficient. The calculation of the sticking coefficient is a challenging task as it is dependent upon the mass of the PAH species to the fourth power [24]. Blanquart [24] also reduced sticking coefficient by a factor of three, in case of napthalene and acenaphthylene, in order to better match with the experimental concentrations of the PAH molecules, signifying the non-linear behaviour. For smaller particles, D'Alessio et al. [25] suggested the dependency is directly proportional to the square of the mass of the particles. The variation of the sticking coefficient with the mass of the precursor species is shown in Fig. 1 and current chosen value is labelled accordingly.

Coagulation

To reduce the complexities in the modeling of particulates, a coalescent coagulation has been assumed, where particles will retain themselves in a spherical shape with increasing diameter after the collision. The coagulation source terms for the n = 0th and 1st moments in the moment transport equation are computed as follows:

$$S_n^{coa.} = -0.5\sum_{i=1}^{\infty}\sum_{j=1}^{\infty}\beta_{i,j}N_iN_j \quad (11)$$

As the total mass of the particles remained unaffected during coagulation process, the coagulation source term for higher moments, i.e. $n \geq 2$ is computed using Eq. (12):

$$S_n^{coa.} = -0.5\sum_{k=1}^{n-1}\binom{n}{k}\sum_{i=1}^{\infty}\sum_{j=1}^{\infty}m_i^n m_j^{n-k}\beta_{i,j}N_iN_j \quad (12)$$

In Eqs. (11), (12), $\beta_{i,j}$ is the collision frequency and is dependent on various coagulation regimes such as continuum, free molecular and transition (intermediate of two). The state of the coagulation process corresponds to their regimes is defined using Knudsen number, i.e. the ratio of the collision diameter to the mean free path. In order to determine the coagulation process regime, closed domain with a fixed

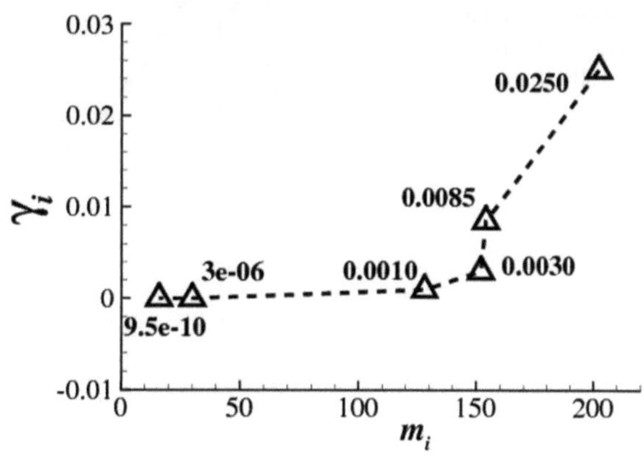

Fig. 1 Variation of the sticking coefficient for the different mass of the precursor species

population of soot particles are considered initially. These particles are initially mono-dispersed and the initial number density and temperature are assigned as 10^{18} m^{-3} and 1500 K. Then, particles are allowed to coagulate and the change in particle number density is tracked over the period of time. Figure 2 shows the evolution of the number density with time and is compared with earlier predictions of Frenklach and Harris [26] stipulating the domination of the free molecular regime.

Surface Growth and Oxidation

The unpretentious soot formed during the nucleation process is followed by the dominant mode of soot formation called surface growth, due to several surface reactions taking place on the surface of the soot particle. However, during the process, the number density of the particles remains consistent, whereas the mass of the soot particles increases substantially. Simultaneously, the soot particles also lose respective mass concentrations due to oxidation in present of O_2 and OH radicals, primarily. The process of surface growth and oxidation are kinetically controlled and involve a higher order of complexities, thus fairly detailed mechanisms are used to approximate these sub-processes such that to enhance the computational efficiency.

2.3 *Soot-Radiation and Soot-Turbulence Interactions*

As discussed previously, the effective absorption coefficient is the summation of the absorption coefficient of soot and the gas as depicted in the following equation:

$$a_{\lambda} = a_i + a_{soot} \tag{13}$$

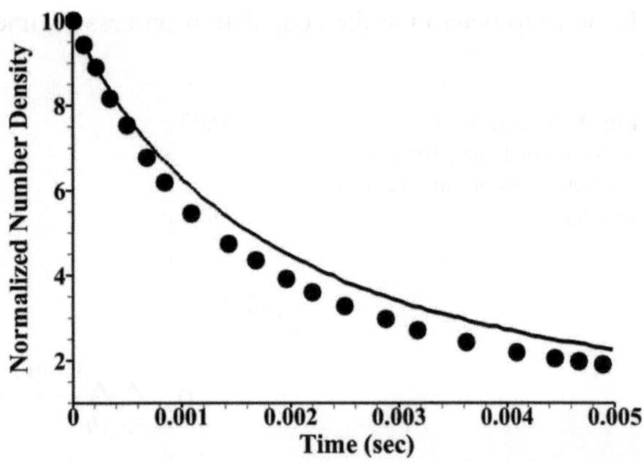

Fig. 2 Evolution of number density for a free molecular regime where the *line* is the current prediction and symbols are results of Frenklach and Harris [26]

The effect of turbulence is included by a single variable PDF in terms of temperature as:

$$\tilde{S}_{soot} = \int \rho S_{soot}(T)P(T)dT \tag{14}$$

The P(T) is the PDF of the variable T and is constructed by solving the soot transport equations. The limits of integration are the maximum and minimum temperature in the flame.

3 Numerical Details

In the present work, the calculations are performed using ANSYS FLUENT 16.0 [27], where the non-gray radiation transfer approach is invoked using a user-defined function (UDF). The second order upwind scheme is used for the discretization of the convective fluxes. SIMPLE algorithm is used for pressure-velocity coupling, while $N_e = 2$ is used for all the MEPDF calculations. In Situ Adaptive Method (ISAT) [28] is used for tabulation of stiff reaction source terms, for all computations an ISAT tolerance of 1e-4 has been used. The weight functions and absorption coefficients as a function of H_2O and CO_2 partial pressures are obtained from Smith [29].

4 Modeling and Boundary Conditions Description

The computational details and imposed boundary conditions used in the present study for comparison with experimental measurements are described in this section.

4.1 Delft Flame III

The first case is a pilot stabilized non-premixed natural gas flame. Peters et al. [30] have described the burner in detail. The fuel jet has a diameter of 6 mm and is surrounded by a rim of 15 mm diameter. Twelve pilot flames of 0.5 mm diameter originate from the rim to stabilize the flame. The rim is surrounded by a concentric circular air inlet of the external diameter of 45 mm.

The whole setup is placed in a wind tunnel and acts as a secondary inlet. Qamar et al. [31] used a sophisticated experimental technique, i.e. Laser Induced Incandescence (LII), to measure the soot production at the downstream locations. Similar to Reddy et al. [32], an axisymmetric 2D grid is considered for the present computational work. A constant mass flow rate issuing from the pilot flame is maintained by adjusting the pilot jet inlet diameter, as suggested by the Naud et al. [33] and Yadav et al. [34]. The computational domain extends 250D in the axial direction and 50D in the radial direction as shown in Fig. 3. Two grids of 400 × 300 and 200 × 150 points are used to study the grid independence and the predictions are shown in Fig. 4. The rest of the computations are performed using the coarse grid (200 × 150). RSM model is used for modeling the underlying turbulent flow, where $C_{\varepsilon 1}$ has been adjusted to 1.6. GRI 1.2 [35] mechanism containing 32 species and 177 reactions is used to represent the chemical kinetics. Experimental boundary conditions are given in Table 1 [36].

4.2 Lifted Turbulent Ethylene/Air Flame

The second case used is a lifted ethylene diffusion flame. The burner details are described by Kohler et al. [15]. The fuel jet has 2 mm internal diameter (D) and 6 mm external diameter. This surrounded by a concentric circular inlet of external diameter 140 mm, acts as an air inlet. The soot measurements in the downstream are performed by Kohler et al. [15]. In similar line with the previous case, also, an axisymmetric 2D grid has been used as with Reddy et al. [37]. The computational the domain used is shown in Fig. 5. Two grids of 720 × 220 and 360 × 110 points are used to study the grid independence and the predictions are shown in Fig. 4; the rest of the computations are performed using the coarse grid (360 × 110).

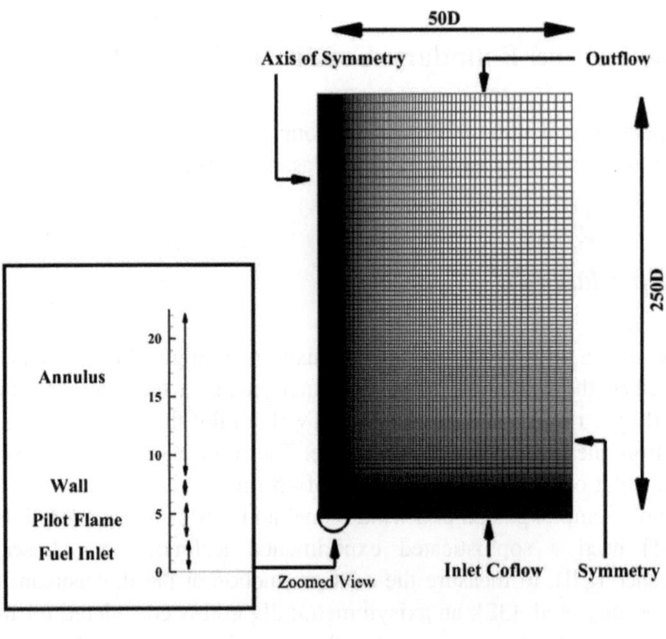

Fig. 3 An illustrative view of the mesh used for Delft Flame III case

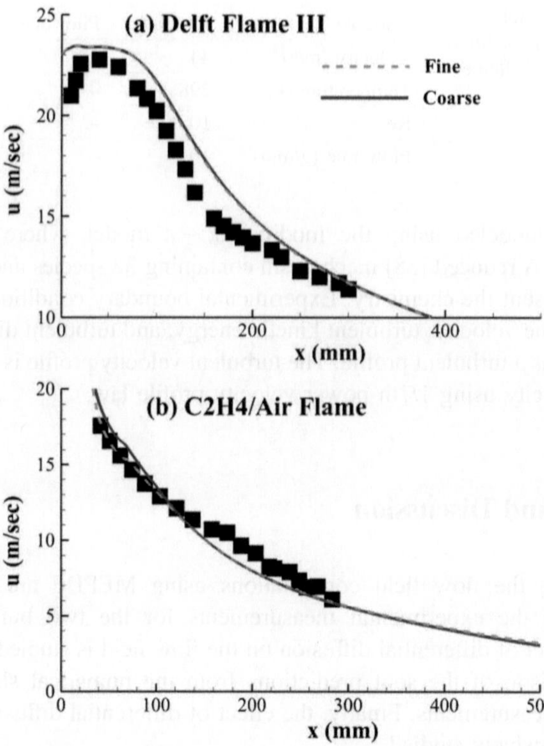

Fig. 4 Centerline distribution of axial velocity: symbols are measurements (Stroomer et al. [36] for Delft Flame III and Kohler et al. [15] for the turbulent lifted flame)

Table 1 Experimental boundary conditions of Delft Flame III

Variable	Jet inlet	Pilot inlet	Coflow Inlet
Velocity (m/s)	21.9	8.34	4.4
Temperature (k)	295	1700	295
Re	9700	–	8800

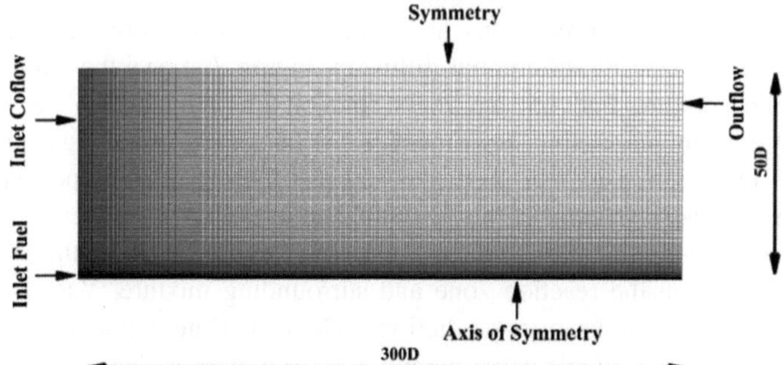

Fig. 5 Illustrative view of the mesh used for turbulent lifted ethylene/air flame

Table 2 Experimental boundary conditions of lifted turbulent ethylene/air flame

Variable	Jet inlet	Pilot inlet	Coflow inlet
Velocity (m/s)	44	–	0.29
Temperature (k)	298	–	312
Re	10,000	–	–
Flow rate (g/min)	10.4	–	320

Turbulence is modeled using the modified $k - \varepsilon$ model, where $C_{\varepsilon 1}$ has been adjusted to 1.6. A reduced [38] mechanism containing 32 species and 206 reactions is used to represent the chemistry. Experimental boundary conditions are given in Table 2 [15]. The velocity, turbulent kinetic energy, and turbulent dissipation at the inlet are given as a turbulent profile. The turbulent velocity profile is provided at the jet inlet of velocity using 1/7th power velocity profile law.

5 Results and Discussion

In this section, the flow field computations using MEPDF are presented and compared with the experimental measurements for the two burners described above. The effect of differential diffusion on the flow field is studied. It is followed by the comparison of the soot predictions from the numerical simulations with experimental measurements. Finally, the effect of differential diffusion on soot has been comprehensively studied.

5.1 Delft Flame III

5.1.1 Structure of the Flame

The calculated radial profiles of the mean axial velocity and temperature are shown in Fig. 6. The mean axial velocity is slightly over predicted near the burner; however, the trend improves along the downstream direction. The radial temperature is also slightly over predicted; this results in a decrease of density and helps in improving the velocity predictions in the downstream direction. The mean axial velocity is unaffected by differential diffusion. Figure 7 shows the axial profile of temperature, mixture fraction and mass fraction of OH radical. The calculated temperature is in reasonable agreement with the experimental data. The radial maximum temperature is over predicted and distribution remains persistent while moving in the radial extents of the domain. The overestimation increases with the inclusion of differential diffusion, and this can be attributed to the decrease in diffusivity between the reaction zone and surrounding mixture. Yadav et al. [34] have also reported similar over-predictions. The centerline temperature is in good agreement with the available experimental data. The over-prediction of temperature

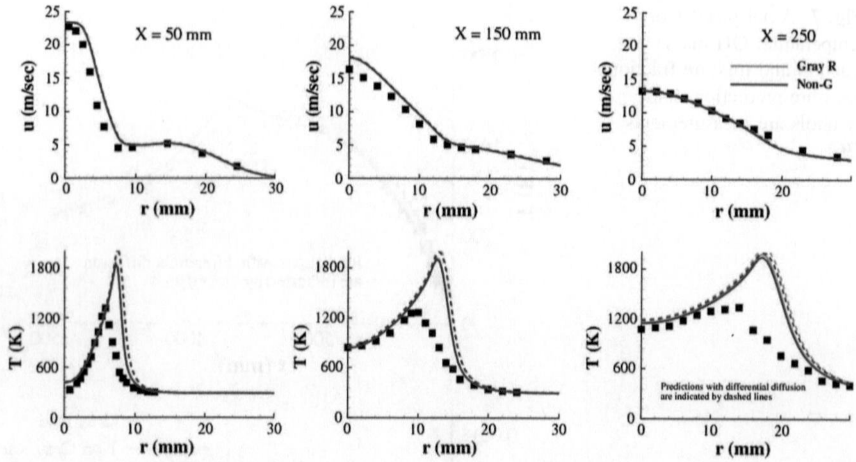

Fig. 6 Radial profiles of axial velocity and temperature at three different locations from the fuel jet exit: *lines* are predictions and symbols are measurements [36]

near the axis is compensated by the overestimation of mixture fraction, resulting in good centreline prediction. Temperature has reduced by ~500 K and stabilized at the maximum temperature of ~1700 K with the inclusion of non-gray radiation and also shifted ~7D in the upward direction. The temperature is shifted further upstream by ~15D with the inclusion of differential diffusion, the maximum temperature is also overestimated by ~50 K. The over-prediction of mixture fraction about the centerline can be attributed to the overprediction of mean velocity at these locations as depicted in Fig. 6.

Since experimental data is only available in the region near to the burner, no precise statement can be made on the length of the flame.

The axial profile of OH mass fraction is shown in Fig. 7. The inclusion of non-gray radiation decreases the mass fraction by a factor of ~2.5. The location of the maximum mass fraction is also shifted upstream with the inclusion of differential diffusion.

5.1.2 Soot Predictions Using Semi-empirical Model

Figure 8 shows the computed axial profile of soot volume fraction using non-gray radiation model in comparison with experimental measurements of Qamar et al. [31] and LES predictions of Mueller et al. [13], Donde et al. [14]. The evolution of the soot volume fraction is observed to occur upstream by the LES predictions in comparison with experimental data. The present approach is able to reasonably predict the location of the maximum soot volume fraction. However, the position shifted upstream by ~7D with the inclusion of differential diffusion. The shift is consistent with the temperature profiles. The magnitude is also increased by a factor

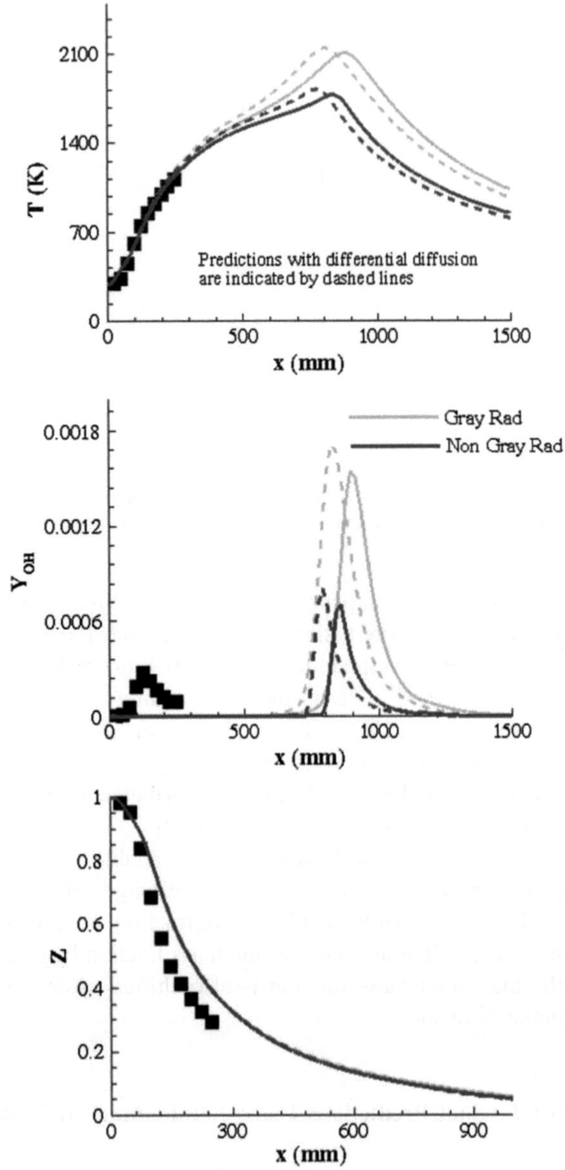

Fig. 7 Axial profiles of temperature, OH mass fraction and mixture fraction: *lines* are predictions and symbols are measurements [36]

of ~1.5, the reason for this behavior is explained subsequently. The decrease in soot volume fraction with the inclusion of the soot-turbulence interaction might be attributed to the reduction of surface growth. When the turbulent fluctuations are combined with the process of surface growth of the soot, these results into the inhibition of the heterogeneous reactions; restricts the number density of the soot

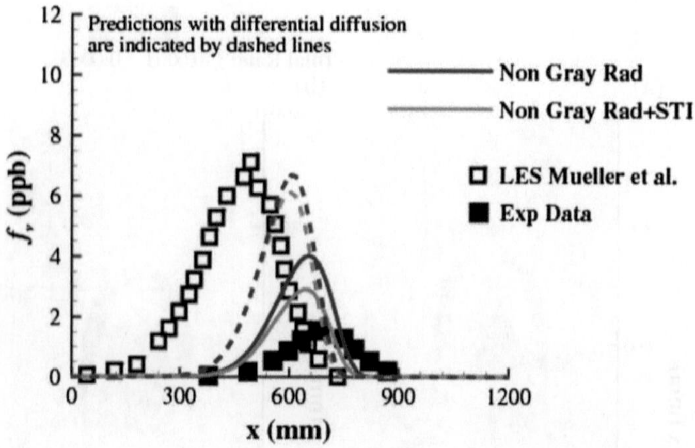

Fig. 8 Axial profiles of soot volume fraction using semi-empirical modeling approach and experimental data [31]

particles. The location of maximum soot volume fraction is not considerably affected by soot-turbulence interaction.

The contour of soot production rate with the inclusion of soot-turbulence interaction is depicted in Fig. 9. The nucleation of soot increases with the inclusion of differential diffusion resulting in an increase of surface growth rate, thereby the increase in net production. The above-mentioned effects result in an increase of soot volume fraction about the centerline. It can also be inferred that the oxidation region is not considerably affected by the differential diffusion.

5.1.3 Soot Predictions Using Method of Moment Model

The sensitivity of the differential diffusion and soot-turbulence interaction along the centerline soot volume fraction distribution is shown in Fig. 10. The downstream shift of the maximum soot volume fraction is observed to be ~4D and ~7D using differential diffusion and unity Lewis number configuration, respectively. However, the peak value of the soot volume fraction is well captured using differential diffusion which can be attributed to the accurate estimation of nucleation zone, whereas a reduction of the factor ~4.4 is observed in the case of unity Lewis number case. After accounting the soot-turbulence interactions in the system, prominent improvement in the soot volume fraction predictions can be seen with the differential diffusion as compared to the unity Lewis number case. This behavior is consistent with the empirical modeling approach. Also, reduction in soot residence time is observed specifically after inclusion of the soot-turbulence interactions. However, the effect of the soot-turbulence fluctuations remains unhindered particularly under the differential diffusion. The location of the nucleation initiation

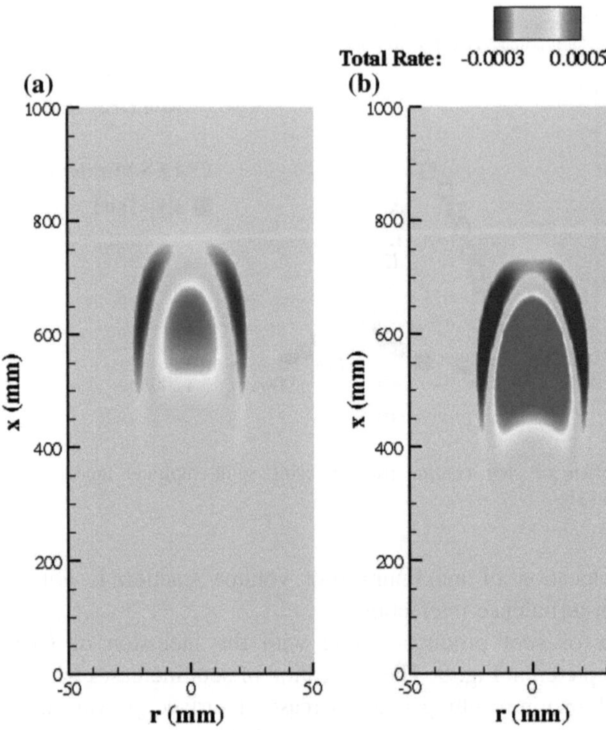

Fig. 9 Contour of soot total rate obtained from the semi-empirical approach with STI where **a** unity Lewis number and **b** differential diffusion

Fig. 10 Axial profiles of soot volume fraction using the method of moment modeling approach and experimental data [31]

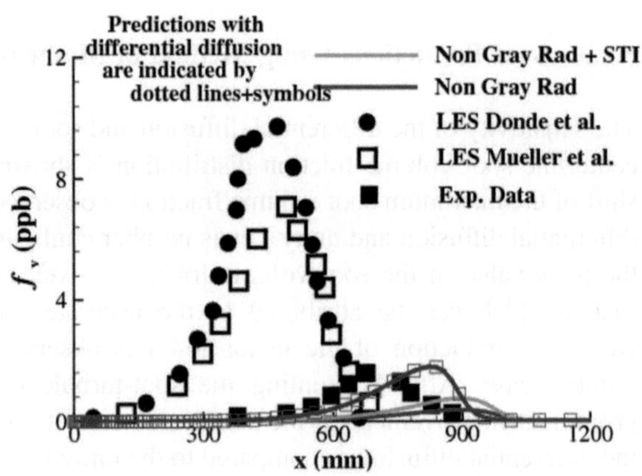

process is well captured by the differential diffusion approach; further predictions are deflected downstream due to the inaccurate predictions of the surface growth process and oxidation due to the OH radicals.

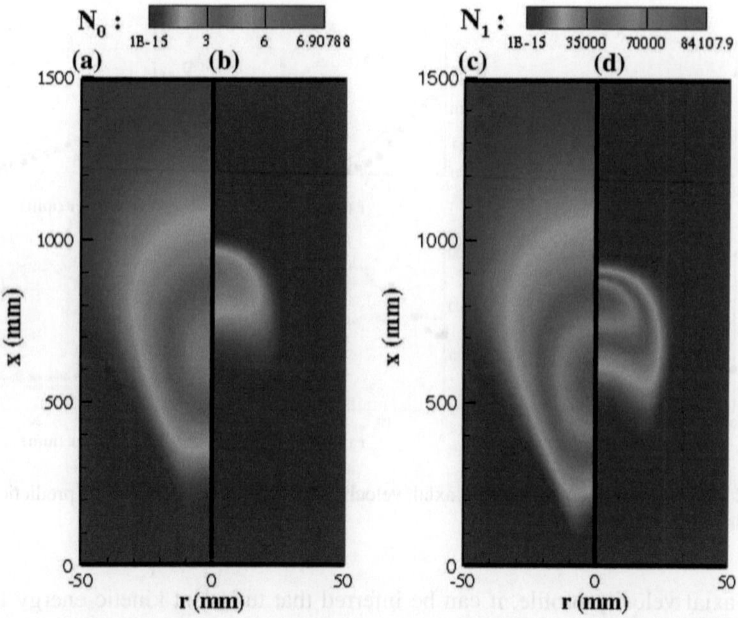

Fig. 11 Contours obtained from the method of moment model, signifying the total particle number density (N_0) and the total mass of the soot particles (N_1) with STI using **a** and **b** unity Lewis number and **c** and **d** differential diffusion

To further understand the effect of the combinations of molecular diffusion, contours of the total particle number density (N_0) and the total mass of the soot particles (N_1) are shown in the Fig. 11. The under-prediction of around 1.5 and 2.5 is observed in the total particle number density, shown in Fig. 11a, c and the total mass of the soot particles, shown in Fig. 11b, d, respectively, in the case of unity Lewis number, resulting in the lesser production of the soot formation. However, the increased predictions along the centerline are observed on account the differential diffusion, resulting in higher soot volume fraction predictions. It can be deduced that unlike empirical modeling approach, the inclusion of differential diffusion and soot-turbulence interactions leading to the satisfactory predictions of the soot volume fraction.

5.2 *Lifted Turbulent Ethylene/Air Flame*

5.2.1 Structure of the Flame

The radial profiles of the mean axial velocity and temperature are shown in Fig. 12. The predictions are in reasonable agreement with the experimental data. From the

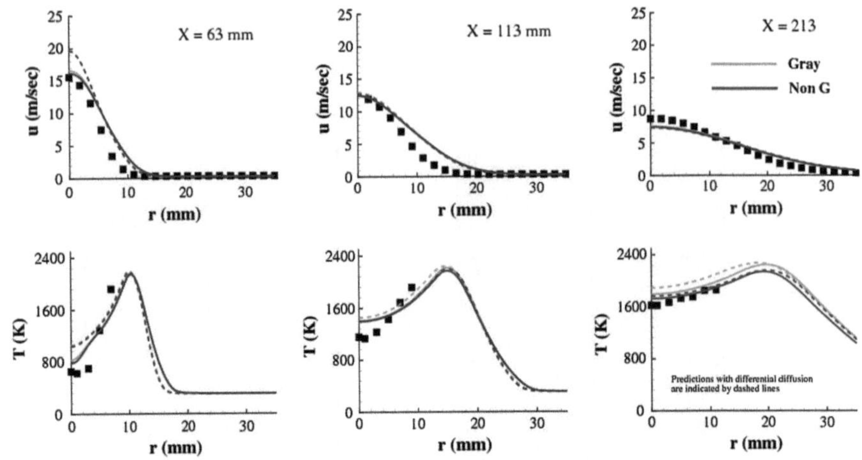

Fig. 12 Radial profiles of the mean axial velocity and temperature: *lines* are predictions and symbols are measurements [15]

mean axial velocity profile, it can be inferred that turbulent kinetic energy is over predicted resulting in an increase of turbulent shear stress causing deceleration of central jet in the further radial extents. The radial velocity predictions of the Kohler et al. [15] are observed to be under-predicted in the downstream locations whereas predictions along the centerline are in good match with the experiments. The lift off height is predicted fairly accurately; experimental lift off height is 22.3 $\pm$ 1.5 mm and predicted value is 22.9 mm. This can be attributed to the MEPDF model which accurately captures the turbulence-chemistry interactions. As mentioned and noted earlier, this particular model uses the advantage of lagrangian PDF approach and is limited to less number of environments which makes it computationally attractive for non-premixed flame computations. The increase in spreading rate results in under-prediction of temperature in the radial direction. As observed for the Delft Flame III, the inclusion of differential diffusion results in the over-prediction of temperature near the centerline. The inclusion of non-gray radiation reduces the centerline temperature by $\sim$300 K and maximum peak temperature limits to 1800 K, further including differential diffusion shifts the maximum flame temperature in the upstream direction by $\sim$15 D and resulting flame height is 415 mm to the experimental measurement of $\sim$410 mm. The reason for the reduction in flame length is same as Delft Flame III.

5.2.2 Soot Predictions Using Semi-empirical Model

The centerline soot volume fraction predictions using non-gray radiation model are reported in Fig. 13. Both the Lee and Fenimore-Jones approach are able to reasonably predict the location of maximum soot volume fraction. The inclusion of

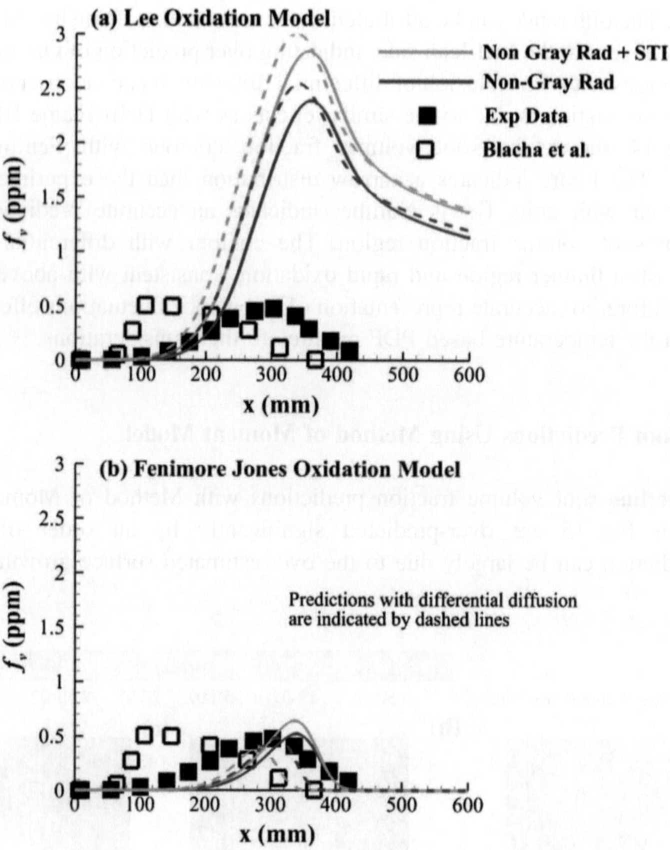

Fig. 13 Axial profiles of soot volume fraction obtained from semi-empirical approach signifying effect of **a** Lee oxidation model and **b** Fenimore-Jones oxidation model: *hollow* symbols corresponds to the sectional model [39] and *Solid* symbols are measurements [15]

differential diffusion results in a shift of location of maximum soot volume fraction in the upstream direction; this is consistent with the temperature predictions and elucidates the impact of flame length on the soot volume fraction distribution. The effect of nucleation on net production is not as noteworthy as the surface growth process. The soot volume fraction by Lee approach is overpredicted by a factor of ~4 compared to the Fenimore-Jones approach. The domination of OH radical compared to molecular oxygen results in a reduction of soot volume fraction with Fenimore-Jones approach, the difference might also be due to the different scaling parameters used for both the approaches and requires further study which is beyond the scope of the current study. The soot volume fraction increases with the inclusion of the soot-turbulence interaction; this can be attributed to the increase in the flame surface area resulting in an increase of net production of soot. Contrarily, the soot volume fraction decreases for the Fenimore-Jones approach with differential diffusion. The difference can be attributed to the reduction of flame height and rapid oxidation of soot in the fuel-lean side, indicating over prediction of OH radical. For the other cases with the inclusion of differential diffusion, it can be observed that the increase in oxidation is due to the similar effects as with Delft Flame III.

Figure 14 shows the soot volume fraction contour with Fenimore-Jones approach. The figure indicates a narrow distribution than the experimental data. The contour with unity Lewis Number indicates an accurate prediction of the maximum soot volume fraction region. The contour with differential diffusion indicates even thinner region and rapid oxidation, consistent with above explanation. Therefore, for accurate representation of turbulence fluctuations effect on soot evolution the temperature based PDF requires further considerations.

5.2.3 Soot Predictions Using Method of Moment Model

The centerline soot volume fraction predictions with Method of Moment model shown in Fig. 15 are over-predicted significantly by an order of 2. This over-prediction can be largely due to the over-estimated surface growth from the

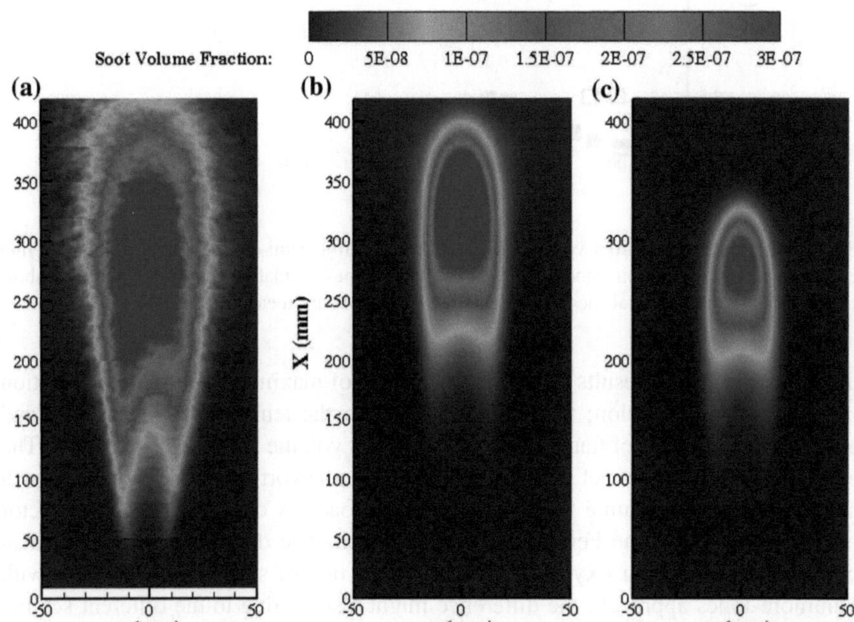

Fig. 14 Contour of soot volume fraction obtained from semi-empirical approach + STI for turbulent lifted flame, signifying **a** experimental [15], **b** unity Lewis number and **c** Differential diffusion

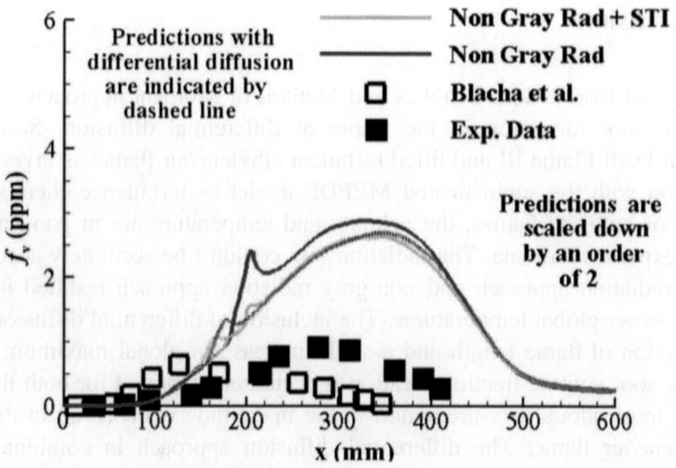

Fig. 15 Axial distribution of soot volume fraction computed using method of moment model: *hollow* symbols corresponds to the sectional model [39] and *solid* symbols are measurements [15]

HACA mechanism. This can also be attributed to the incapability of acetylene as the only soot precursor, resulting into the inaccurate computation of nucleation process. Under non-gray radiation, the soot volume fraction distribution with differential diffusion reduced by merely a factor of ~ 1.1 as compared to the unity Lewis number case. A kink is observed in the soot volume fraction distribution, which is particularly resulting from locally enhanced nucleation occurring due to an over-predicted temperature in the region, which can be seen in the radial distributions of the temperature. However, a similar phenomenon is not observed in the case of the empirical modeling approach. Further, this kink is depicting decaying phenomenon with the inclusion of the differential diffusion. Unlike empirical modeling approach, the marginal shift is observed in the soot volume fraction and its distribution largely remains unaffected after the inclusion of the soot-turbulence interactions. The predicted location of the maximum soot volume fraction is not well captured; a noticeable difference of 35D in the downstream direction is observed. This can be attributed to the spatially inaccurate nucleation and surface growth phenomenon predicted by the Method of Moment model. Overall extent or surface area of the soot volume fraction distribution is also over-predicted, primarily due to the large residence time associated with the soot evolution processes. This clearly elucidates, firstly, other soot precursor species should be included in order to accurately predict the soot concentration developed from nucleation phenomena and secondly, parameters of the soot oxidation and surface growth should be tuned up such that to obtain near accuracy in the soot volume fraction predictions.

6 Conclusion

In the present work, Moss-Brookes and Method of Moment approach are used to investigate soot formation in the frame of differential diffusion. Soot volume fraction in Delft Flame III and lifted turbulent ethylene/air flames is investigated in conjunction with the sophisticated MEPDF model as turbulence-chemistry interactions. For both the flames, the velocity and temperature are in good agreement with the experimental data. The radiation loss couldn't be accurately accounted by the gray radiation approach and non-gray radiation approach resulted in the significantly lower global temperatures. The inclusion of differential diffusion resulted in a reduction of flame length and a slight increase in global maximum temperatures. The soot volume fraction location is accurately captured for both the flames, however, tremendous over-prediction in the magnitude is observed in the case of the ethylene/air flame. The differential diffusion approach in combination with semi-empirical model resulted in the spatial variation of the maximum soot volume fraction location in the upstream direction and a net increase in soot production, whereas the spatially downstream evolution of the soot volume fraction is observed with Method of Moment model. With the semi-empirical model, soot-turbulence interactions resulted in a decrease of soot volume fraction in Delft Flame III and increase in ethylene/air flame except for Fenimore-Jones approach with differential diffusion. In contrast, soot evolution is increased in Delft Flame III and marginally decreased in case of ethylene/air flame. The results elucidate the complex nature of soot-turbulence-chemistry interactions, indicating the need for modeling with representative considerations and further comment tuning of the surface growth and oxidation model parameters such that to capture accurate soot evolution in the ethylene/air flame.

References

1. Wang L (2004) Detailed chemistry, soot, and radiation calculations in turbulent reacting flows. Dissertation. The Pennsylvania State Univeristy
2. Haynes B, Wagner H (1981) Soot formation. Prog Energy Combust Sci 7:229–273
3. Bockhorn H (1994) Soot formation in combustion. Springer
4. Kennedy I (1997) Models of soot formation and oxidation. Prog Energy Combust Sci 23:95–132
5. Nilsen V, Kosály G (1999) Differential diffusion in turbulent reacting flows. Combust Flame 117(3):493–513
6. Mardani A, Tabejamaat S, Ghamari M (2010) Numerical study of influence of molecular diffusion in the mild combustion regime. Combust Theor Model 14(5):747–774
7. Richardson ES, Chen JH (2012) Application of PDF mixing models to premixed flames with differential diffusion. Combust Flame 159(7):2398–2414
8. De A, Dongre A (2015) Assessment of turbulence-chemistry interaction models in mild combustion regime. Flow Turbul Combust 94(2):439–478
9. Navarro-Martinez S, Rigopoulos S (2012) Differential diffusion modelling in LES with RCCE-reduced chemistry. Flow Turbul Combust 89(2):311–328

10. Pitsch H, Riesmeier E, Peters N (2000) Unsteady flamelet modeling of soot formation in turbulent diffusion flames. Combust Sci Technol 158:389–406
11. Kronenburg A, Bilger R (1997) Modelling of differential diffusion in nonpremixed nonreacting turbulent flows. Phys Fluid 9(5):1435–1447
12. Wang H (2016) Consistent flamelet modelling of differential molecular diffusion for turbulent non-premixed flames. Phys Fluids 28(3):035102
13. Mueller ME, Pitsch H (2012) LES model for sooting turbulent nonpremixed flames. Combust Flame 159(6):2166–2180
14. Donde P, Raman V, Mueller ME, Pitsch H (2013) LES/PDF based modeling of soot–turbulence interactions in turbulent flames. Proc Combust Inst 34(1):1183–1192
15. Köhler M, Geigle K-P, Blacha T, Gerlinger P, Meier W (2012) Experimental characterization and numerical simulation of a sooting lifted turbulent jet diffusion flame. Combust Flame 159 (8):2620–2635
16. Lignell DO, Chen JH, Smith PJ, Lu T, Law CK (2007) The effect of flame structure on soot formation and transport in turbulent nonpremixed flames using direct numerical simulation. Combust Flame 151(1–2):2–28
17. McGee HA (1991) Molecular engineering. MaGraw-Hill, New York
18. Siegel R (2001) Thermal radiation heat transfer. CRC Press
19. Yadav R, Kushari A, Verma AK, Eswaran V (2013) Weighted sum of gray gas modeling for nongray radiation in combusting environment using the hybrid solution methodology. Numer Heat Transf Part B Fundam 64(2):174–197
20. Brookes S, Moss J (1999) Predictions of soot and thermal radiation properties in confined turbulent jet diffusion flames. Combust Flame 116(4):486–503
21. Reddy M, De A, Yadav R (2015) Effect of precursors and radiation on soot formation in turbulent diffusion flame. Fuel 148:58–72
22. Frenklach M (2002) Method of moments with interpolative closure. Chem Eng Sci 57 (12):2229–2239
23. Frenklach M (2002) Reaction mechanism of soot formation in flames. Phys Chem Chem Phys 4(11):2028–2037
24. Blanquart G (2008) Chemical and statistical soot modelling. Doctoral dissertation, Stanford University
25. D'Alessio A, Barone AC, Cau R, D'Anna A, Minutolo P (2005) Surface deposition and coagulation efficiency of combustion generated nanoparticles in the size range from 1 to 10 nm. Proc Combust Inst 30(2):2595–2603
26. Frenklach M, Harris SJ (1987) Aerosol dynamics modelling using the method of moments. J Colloid Interf Sci 118(1):252–261
27. Ansys Fluent 16.0 User's guide, Csnosburg. PA, USA
28. Pope SB (1997) Computationally efficient implementation of combustion chemistry using in situ adaptive tabulation. Combust Theory Model 1(1):41–63
29. Smith TF, Shen ZF, Friedman JN (1982) Evaluation of coefficients for the weighted sum of gray gases model. J Heat Transfer 104(4):602–608
30. Peeters TWJ, Stroomer PPJ, de Vries JE, Roekaerts DJEM, Hoogendoorn CJ (1994) Comparative experimental and numerical investigation of a piloted turbulent natural-gas diffusion flame. Symp Combust 25(1):1241–1248
31. Qamar NH, Alwahabi ZT, Chan QN, Nathan GJ, Roekaerts D, King KD (2009) Soot volume fraction in a piloted turbulent jet non-premixed flame of natural gas. Combust Flame 156 (7):1339–1347
32. Manedhar Reddy BB, De A, Yadav R (2015) Numerical investigation of soot formation in turbulent diffusion flame with strong turbulence–chemistry interaction. ASME J Thermal Sci Eng Appl 8(1)
33. Naud B, Jimenez C, Merci, B, Roekaerts D (2007) Transported PDF calculations of the piloted jet diffusion flame 'delft flame III' with complex chemistry: study of the pilot model. Computational Combustion, ECCOMAS Thematic Conference, Delft

34. Yadav R, Kushari A, Eswaran V, Verma AK (2013) A numerical investigation of the eulerian PDF transport approach for modeling of turbulent non-premixed pilot stabilized flames. Combust Flame 160(3):618–634
35. Smith GP, Golden DM, Frenklach M, Moriarty NW, Eiteneer B, Goldenberg M, Bowman CT, Hanson RK, Song S, Gardiner WC, Lissianski VV "No Title". http://www.me.berkeley.edu/gri_mech/
36. Stroomer PPJ (1995) Turbulence and OH structures in flames. Delft University of Technology
37. Busupally MR, De A (2016) Numerical modeling of soot formation in a turbulent C2H4/air diffusion flame. Int J Spray Combust Dyn 8(2):67–85
38. Luo Z, Yoo CS, Richardson ES, Chen JH, Law CK, Lu T (2012) Chemical explosive mode analysis for a turbulent lifted ethylene jet flame in highly-heated coflow. Combust Flame 159(1):265–274
39. Blacha T, Di Domenico, M, Köhler M, Gerlinger P, Aigner M (2011) Soot modeling in a turbulent unconfined C2H4/air jet flame. In 49th AIAA aerospace science meeting including new horizons forum aerospace exposition, pp 1–10

Development of a Mobile Emission Test Car for Indian Railways

Anirudh Gautam, Manish Agarwal and Mohd Amil

Abstract In a historic judgment the National Greens Tribunal of India has stipulated that Indian Railways should specify emission standards for diesel locomotives of Indian Railways and implement technologies to reduce these pollutants from diesel locomotives in a time bound manner. In order to decide on the emission standards, it is essential that an emission inventory of the diesel locomotives be obtained from across different areas, working environments, different configurations and age groups. Indian Railways has a population of about 5000 diesel locomotives and 56 diesel sheds and it is not possible to establish mass emission measurement systems at all diesel sheds because of the high cost of these measurement systems as well as a scientific group of personnel required for carrying out these measurements as per International standards like US EPA and Europe UIC. Both of these international emission standards stipulate mass based measurements of the pollutants in order to arrive at the emission levels of a diesel locomotive. Engine Development Directorate at Research Designs & Standards Organisation (RDSO) has setup mass emission measurement systems at RDSO for measuring emissions from test engines. At least, 100 diesel locomotives are required to be measured from different areas, configurations, age groups, etc. in order to arrive at statistically significant upper and lower control limits. It is not physically possible to bring these 100 locomotives to Engine Development Directorate at RDSO, therefore a mobile emissions test car was designed and manufactured by Indian Railways. This mobile emission test car is capable of measuring mass emissions from the diesel locomotives as per the International standards. This chapter discusses the process of how this mobile emission test car was envisaged and manufactured. This included the layout, selection of the equipments and ease of operation so as to meet the international emission measurement standards.

A. Gautam (✉) · M. Agarwal · M. Amil
Ministry of Railways, RDSO Lucknow, India
e-mail: gautam.anirudh@gmail.com

A.K. Agarwal et al. (eds.), *Locomotives and Rail Road Transportation*,
DOI 10.1007/978-981-10-3788-7_11

1 Introduction

Indian Railways has a fleet of about 5000 locomotives maintained in about 50 diesel locomotive maintenance shed, which run all across the country [1]. Although these diesel locomotives consume only about 4% of the total diesel fuel consumed in India, hence they do not contribute significantly in terms of polluting the atmosphere, yet there has been growing concerns (NO_x, PM, CO, CO_2) to regulate the emissions from these locomotives. Emission limits for few regulated pollutants from diesel locomotives have been stipulated in only North America and Europe and other countries are considering similar emission standards. In India, the National Green Tribunal (NGT) [2], in response to a Public Interest Litigation (PIL) mandated that Indian Railways should also fix emission standards for its diesel locomotives and plan a roadmap for reducing these so that International emission standards are met.

Research Designs & Standards Organisation (RDSO) under the Ministry of Railways (MoR) is the central agency for carrying out Railway research in India. Under RDSO, the Engine Development Directorate (EDD) has set-up facilities for carrying out engine research and development. In EDD, equipments have been installed for measuring mass emissions from locomotive diesel engines as per International standards like ISO 8178 and US CFR. For formulating the emission standards of the diesel locomotives in India a sample of about 100 such locomotives with different configurations, areas of running, age, power etc. are required. For operational reasons it is not possible to move these locomotives to the EDD for mass emission measurement. Hence a need for a mobile emission test car which can measure mass emissions as per International standards was strongly felt. In addition, after the up gradation of technology to reduce the pollutants level a certification process must be in place for field monitoring of the fleet. Such a mobile emission test car had not been built anywhere in the world, although General Electric (GE) has built a mobile coach for carrying of the emission equipments. Another important objective for design and development of the mobile Emission Test Car is the high altitude certification of the diesel locomotives as well save fuel and the environment.

1.1 Emission Standards for Diesel Locomotives

As mentioned earlier, USA and Europe are the only regions in the world which have promulgated emission norms for diesel locomotives. Other countries like India, Australia, Japan are in the process of framing them. The emission legislation in Europe and USA are discussed below [3–5].

1.1.1 Europe

European emission standards have been framed for non-road mobile machinery with progressively stringent tiers called a Stage I–V standards. The main stages of EU non-road emission standards are:

- Stage I/II Stage I emission standard was brought into effect in the year 1997 (implemented in 1999) and stage II implemented during 2001–2004 which was on the basis of the engine power output. Locomotive, aircraft and ship engines were not covered by the Stage I/II standards.
- Stage III/IV-emission standards were adopted in the year 2004. Stage III and IV standards expanded the scope of implementation to locomotive and marine engines. Stage III/IV legislation are applicable to new vehicles and equipment; replacement engines to be used in machinery already in use with the exception of railcar, locomotive and inland waterway vessel propulsion engines are required to in line with the limits when the equipment was originally placed on the market.
- Stage V emission regulations were proposed in the year 2014. There are various categories of engines which are included in the Stage V emission standards. Amongst these the category RLL is for engines for the propulsion of railway locomotives and category RLR is for Engines for the propulsion of railcars; Another notable feature of stage V is the adoption of particle number (PN) emission limits for different categories of CI engines.

Rail Traction Engines

Table 1 lists the emission values for stage IIIA and IIIB standards. There are no Stage IV standards for rail traction engines.

From Table 1 it can be inferred that from stage IIIA to IIIB, the PM has reduced by an order of magnitude, NO_x emissions have remained more or less constant, HC emissions have decreased three times and the CO emission values have remained the same.

Table 1 Stage III A/B emission standards for rail traction engines [6]

Category	Net power	Year	CO	HC	HC + NO_x	NO_x	PM
	kW		g/kWh				
Stage III A							
RC A	P > 130	2006	3.5	–	4.0	–	0.2
RL A	130 ≤ P ≤ 560	2007	3.5	–	4.0	–	0.2
RH A	P > 560	2009	3.5	0.5*	–	6.0*	0.2
Stage III B							
RC B	P > 130	2012	3.5	0.19	–	2.0	0.025
R B	P > 130	2012	3.5	–	4.0	–	0.025

*HC = 0.4 g/kWh and NO_x = 7.4 g/kWh for engines of P > 2000 kW and D > 5 L/cylinder

Proposed Stage V emission standards for rail traction engines are shown in Table 2. These are applicable to engines of any power rating and any type of ignition.

From Table 2 it is seen that while other emission limits remain the same the requirement of particle number has been introduced in stage V of the EU emission regulation.

1.1.2 USA

- US emission standards for railway locomotives are applicable to newly manufactured and re-manufactured locomotives. In all there are Tier 0 to Tier 4 standards which progressively set lower limits for the emission of regulated pollutants. These are explained below:
- Tier 0–2 standards were adopted in the year 1997 and became applicable from the year 2000. These standards are applicable to the locomotives manufactured from 1973 during new manufacture or remanufacture. These standards are expected to be met though engine design modifications without the use of exhaust after-treatment devices.
- Tier 3–4 standards were adopted in 2008 (implemented 2011/12). Tier 3 standards are expected to be complied by engine modifications whereas Tier 4 standards may require after-treatment devices and applicable from the year 2015. Another noteworthy feature of the 2008 regulation is that it has introduced more stringent emission standards for re-manufactured Tier 0–2 locomotives.

The Tier 0–2 emission standards are shown in Table 3. A dual cycle approach has been considered in the regulation, i.e., all locomotives are required to comply with both the line-haul and switch duty cycle standards, regardless of intended usage.

Locomotive engines are required to meet smoke opacity standards (Table 4).

Tier 3–4 Standards

The 2008 regulation has made the Tier 0–2 standards more stringent for existing locomotives, as well as provided new Tier 3 and Tier 4 emission standards:

Table 2 Proposed stage V emission standards for rail traction engines

Category	Net power	Date	CO	HC[a]	NO_x	PM	PN
	kW		g/kWh				1/kWh
RLL-v/c-1 (locomotives)	P > 0	2021	3.50	4.00[b]		0.025	–
RLR-v/c-1 (railcars)	P > 0	2021	3.50	0.19	2.00	0.015	1×10^{12}

[a]A = 6.00 for gas engines
[b]HC + NO_x

Table 3 Tier 0–2 locomotive emission standards, g/bhp h [6]

Duty cycle	HC*	CO	NO_x	PM
Tier 0 (1973–2001)				
Line-Haul	1.0	5.0	9.5	0.60
Switch	2.1	8.0	14.0	0.72
Tier 1 (2002–2004)				
Line-Haul	0.55	2.2	7.4	0.45
Switch	1.2	2.5	11.0	0.54
Tier 2 (2005 and later)				
Line-Haul	0.3	1.5	5.5	0.20
Switch	0.6	2.4	8.1	0.24
Non-regulated locomotives (1997 estimates)				
Line-Haul	0.5	1.5	13.5	0.34
Switch	1.1	2.4	19.8	0.41

*HC standard is in the form of THC for diesel engines

Table 4 Locomotive smoke standards, % opacity (normalized)

	Steady-state	30-s peak	3-s peak
Tier 0	30	40	50
Tier 1	25	40	50
Tier 2 and later	20	40	50

- Tier 0–2 standards—More stringent emission standards for existing locomotives when they are remanufactured,
- Tier 3 standards are for newly-built and re-manufactured locomotives for near term.
- Tier 4 standards are for longer-term standards for newly-built and re-manufactured locomotives.

Excluded are the (1) historic steam-powered locomotives, (2) electric locomotives, and (3) some existing locomotives owned by small businesses. Engines used in locomotive-type vehicles with less than 750 kW total power, engines used only for hotel power, and engines used in self-propelled passenger-carrying railcars. The engines used in these smaller locomotive-type vehicles are regulated by the nonroad engine requirements. These emission standards are shown in Tables 5 and 6.

Tier 3–4 locomotives must also meet smoke opacity standards as specified in Table 4.

From Tables 5 and 6 it can be observed that the emission limits have been made narrower and that is no synchronization between the US and the EU diesel locomotive standards.

Table 5 Line-Haul locomotive emission standards (g/bhp h) www.dieselnet.com

Tier	Manufacturing year	Date	HC	CO	NO_x	PM
Tier 0[a]	1973–1992[c]	2010[d]	1.00	5.0	8.0	0.22
Tier 1[a]	1993[c]–2004	2010[d]	0.55	2.2	7.4	0.22
Tier 2[a]	2005–2011	2010[d]	0.30	1.5	5.5	0.10[e]
Tier 3[b]	2012–2014	2012	0.30	1.5	5.5	0.10
Tier 4	2015 or later	2015	0.14[f]	1.5	1.3[f]	0.03

[a]Tier 0–2 line-haul locomotives must also meet switch standards of the same tier
[b]Tier 3 line-haul locomotives must also meet Tier 2 switch standards
[c]1993–2001 locomotive that was not equipped with an intake air coolant system are subject to Tier 0 rather than Tier 1 standards
[d]As early as 2008 if approved engine upgrade kits become available
[e]0.20 g/bhp h until January 1, 2013 (with some exceptions)
[f]Manufacturers may elect to meet a combined NO_x + HC standard of 1.4 g/bhp h

Table 6 Switch locomotive emission standards, g/bhp h [6]

Tier	MY	Date	HC	CO	NO_x	PM
Tier 0	1973–2001	2010[b]	2.10	8.0	11.8	0.26
Tier 1[a]	2002–2004	2010[b]	1.20	2.5	11.0	0.26
Tier 2[a]	2005–2010	2010[b]	0.60	2.4	8.1	0.13[c]
Tier 3	2011–2014	2011	0.60	2.4	5.0	0.10
Tier 4	2015 or later	2015	0.14[d]	2.4	1.3[d]	0.03

[a]Tier 1–2 switch locomotives must also meet line-haul standards of the same tier
[b]As early as 2008 if approved engine upgrade kits become available
[c]0.24 g/bhp h until January 1, 2013 (with some exceptions)
[d]Manufacturers may elect to meet a combined NO_x + HC standard of 1.3 g/bhp h

2 Development of Clean Diesel Technology and Its Field Monitoring: A Prime Objective for the Development of ETC

EDD has done a project for developing a roadmap for evolving cleaner diesel technologies for its fleet of diesel locomotives. This R&D project was carried out in association with a leading engine design consultant. As part of this project a 1-D model of the diesel locomotive engine was developed on Ricardo WAVE software and then various permutations and combinations were tried to ascertain how the engine configuration could be upgraded to meet International Emission Standards. US EPA Tier 2 standards for diesel locomotives were targeted. Engine cylinder head ports were evaluated for their efficiency and other engine components were analysed for their capacity to bear higher loads and thermal conditions. As a result of this study following roadmap emerged for making the engine emission standards compliant (Table 7). The functional specification primarily aimed at the injection timing, Mean Effective Injection pressure (MEIP) increase, stiffer cam rate, recommendation for final combustion chamber detail design and interaction with nozzle cone angle (done in a separate project for development of CReDI for the locomotive engine), number and size of the injector holes, increase in the compression ratio, increase in the trapped air fuel ratio (TAFR) by valve overlap increase and boost pressure increase.

Table 7 Recommended changes to the diesel locomotive engine for emission standards compliance

Attribute	Value	Comment
Injection timing	−7.5° ATDC (Mode 1 and 2	Appropriate for Tier 2 NO_x targets of 6.25 g/kWh (Mode 1) and 9 g/kWh (Mode 2)
MEIP	725 bar (prime path) 650 bar (alternative to be investigated)	Pressure increase to address particulate emissions. Optimum NO_x-PM trade-off may be found with higher MEIP/camrate and retarded timing, or lower MEIP rate and more advanced timing. Investigation recommended
Fuel cam rate	0.7 mm/deg (prime path) 0.65 mm/deg (alternative to be investigated) Or: CRS	Common rail gives future proof injection flexibility and may be engineered to reduce stress on crankcase
Nozzle cone angle	TBC	Recommend final combustion chamber detail design and interaction with nozzle cone angle is investigated
Injector nozzle	9 × 0.304 mm	Based on DIPEN analysis to give target values
Compression ratio	14:1	Analysis shows optimum BSFC at 14:1
Trapped AFR	25:1 minimum at Mode 2 32:1 at Mode 1	Total AFR increased to 27:1 at Mode 2 via valve overlap reduction and boost pressure increase. Required to address Mode 2 particulate emissions
Boost pressure	1.15 bar at Mode 2 2.7 bar at Mode 1 Assuming recommended charge air temperature is achieved	Unchanged at Mode 2 provided TAFR can be increased by reducing valve overlap. Reduction in turbine flow to ~92% necessary to compensate for higher compression/expansion ratio, the smaller turbine results in higher boost pressure at Mode 1. Alternative, VGT should improve AFR control across

(continued)

Table 7 (continued)

Attribute	Value	Comment
		speed range and allow optimum BSFC under all ambient conditions
Boost (charge air temperature)	As low as possible at Mode 1 and 2. Coolant temperature of 30 °C with charge cooler effectiveness of 81%	Recommend cooler group with separate LT water circuit or air/air charge cooler. Temperature management (e.g. shutters, fan control) for light load, especially with low valve overlap
Valve overlap	Reduce to 95–100° by advancing EVC	To improve TAFR, final choice to be validated at light load
Pmax	118 bar	No increase over baseline, although compression ratio increases over baseline, retarded timing offsets this. Protect for 140 bar for uprate potential
Exhaust temperature	610 °C port/560 °C turbine	Confirm with turbocharger supplier
Intake port flow	Cf > 0.61 @ 0.3 L/d	Develop port and rectify faults identified

The boost pressure increase is through the use of Variable Geometry Turbine (VGT) turbocharger, maintaining the peak cylinder pressure to 118 bar, higher exhaust gas temperatures and higher exhaust energies and increase in the intake flow coefficients of the intake ports. To test the effectiveness of these changes (as and when implemented), will require a mobile emission test car to verify and reiterate where necessary.

2.1 Design Layout of the Emission Test Car

As a first step a dialogue was initiated with Integral Coach factory (ICF) which manufactures passenger coaches for Indian Railways. It was the considered opinion of ICF designers that the coach must be fitted with air springs and special buff gear to minimize the effect of the shocks and vibrations that a coach encounters during its run on the railway track. A wide body of the coach was decided upon which would not infringe the maximum moving dimensions of the IR. Therefore a LHB design shell with air springs fitted ICF bogie were decided for manufacture of the

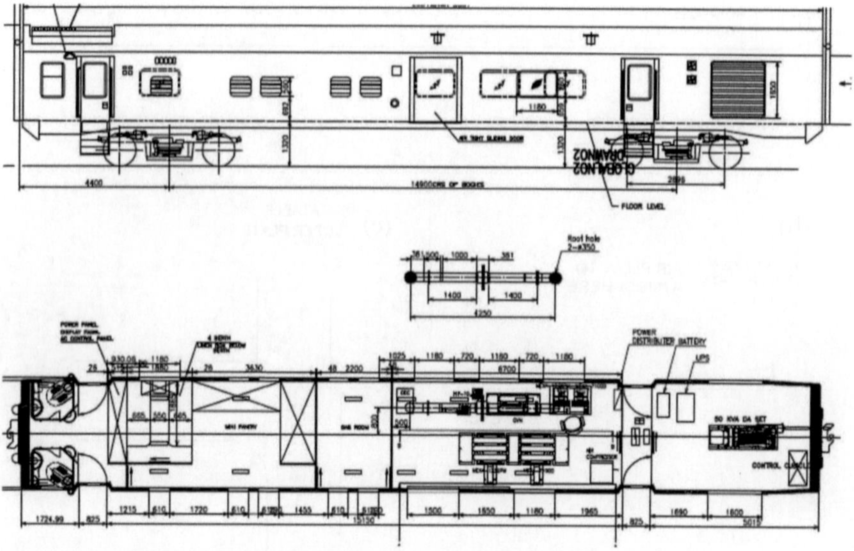

Fig. 1 Layout of the emission test car with various equipments

Emission Test car. It was also deliberated upon whether the mobile emission test car be self-propelled or not. Chief Mechanical Engineer of ICF (later Member Mechanical at Railway Board) was of the view it will not be possible to carry a self-propelled car along with a train without imposing speed restriction. Therefore a non-self-propelled Mobile Emission Test car was decided upon as the final choice. Final layout of the Emission Test Car is show in Figs. 1.

ISO 8178 is a collection of documents for defining emission standards for non-road engines in the European Union, United States, Japan and other countries. In the ISO 8178 there is a restriction on the maximum distance between the point of input of the exhaust sample and the point of input into the measurement equipment. It is also required that the flow of the exhaust be laminar at the point of input to the exhaust measurement equipment. Therefore the very first challenge was the design of the exhaust tube from exhaust manifold of the engine to the emission test car. Various combinations were considered which are shown in Fig. 2.

It was finally decided to have the type of arrangement as shown in Fig. 2e. This allowed the smooth flow of exhaust into the ETC without undue increase in the flow pressure due to flap type relief valve. The equipment used in the ETC are listed in Table 8.

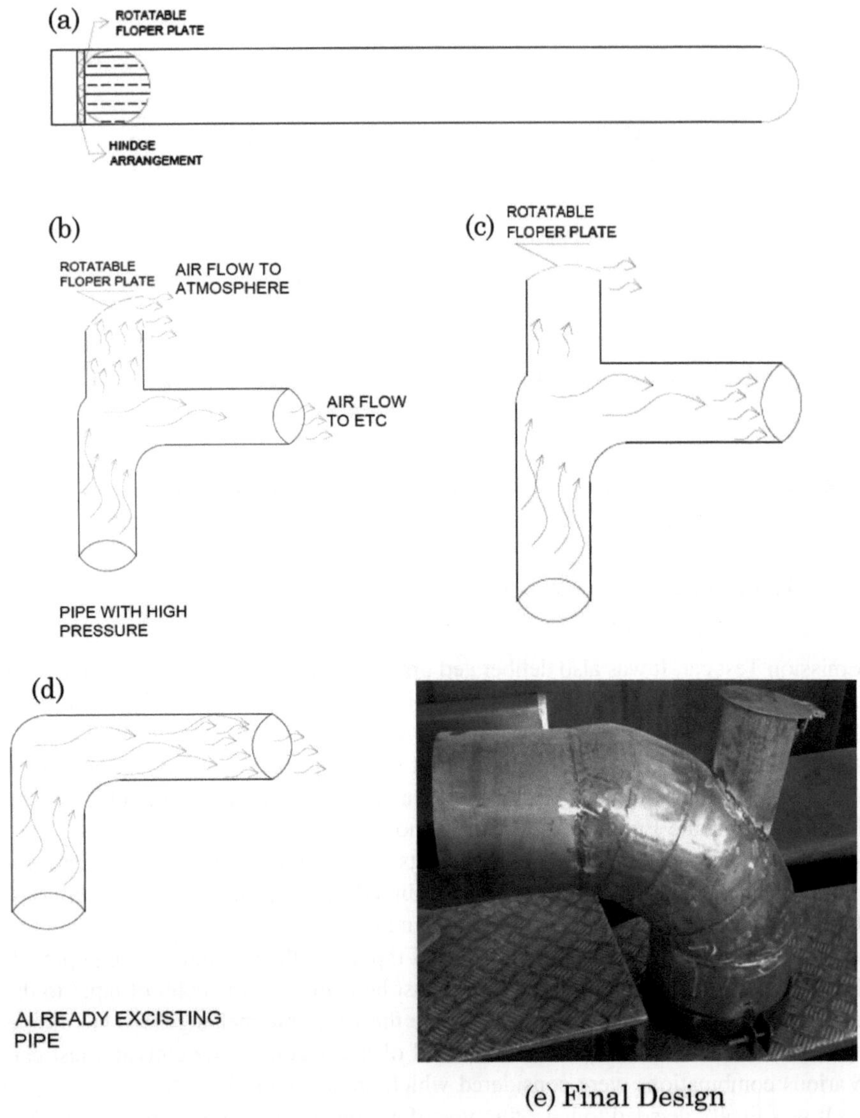

Fig. 2 Various combinations of inlet pipe considered

Table 8 List of all equipments used in the ETC

Utility list								
No.	Equipment weight (kg)	Location of equipment and power supply	Electric (kVA) frequency 50 Hz ± 1.5 Hz	Heat load (kJ/H)	Drain	Factory air (Φ 6/4 Nylon Tube Union)	Purge by air	Exhaust open air
			AC230 V 1Φ					
1	UPS		(16.5 (100 A))					
2	MEXA-7100D (MCU) 350kq		2(30 A)	2160	Required (plate)			6 L/min (open air) (26/19Tube)
3	OVN-72X 120 kg		4(30 A))	4321		400–600 kPa	Purge by AIR (oil, dust free) 20L/min when purge	10 L/min (open air) (26/19Tube)
4	MEXA-1230PM (MCU) 350kq		5(30 A)	5401	Required (PLale)	300–500 kPa	Purge by AIR (oil, dustfree) 10 L/min when purge	10 L/min (open air) (26/19 Tube)
6	Opacity meter		0.5(10 A)	500		34.5 kPa	85 L/min	
6	Compressor		3(20 A)	3240				
7	G0C-703 25 kg		2(10 A)	2160				
Total			16.5 kVA(100 17782(kJ/H)					

2.2 Layout of Exhaust Manifold

The next consideration was the layout of the exhaust manifold. Two openings had to be provided in the roof of the car and a suction pump at the end of the pipe for drawing the exhaust gases into the exhaust manifold and to maintain a steady and laminar flow of the exhaust gases. An inside view of the final arrangement is shown in Fig. 3.

Figure 4 presents the outline view of the sampling exhaust pipe. Further exhaust sampling tubes are attached to the gaseous emission bench and the online real time particulate emission equipment. Both benches were located in specially designed and fabricated frameworks for reduction of vibrations and shocks to accepted levels for the scientific grade equipments. Metallic wire rope springs were used to further filter out the vertical and horizontal vibrations.

2.3 Fuel Measurement System

Fuel measurement is required to calculate the Air Fuel ratio (A/F ratio) which is turn is required to calculate the emission values in g/kWh. The gaseous emissions are

Fig. 3 Inside layout view of the exhaust sampling pipe

Fig. 4 Outline layout view of the sampling exhaust pipe [7]

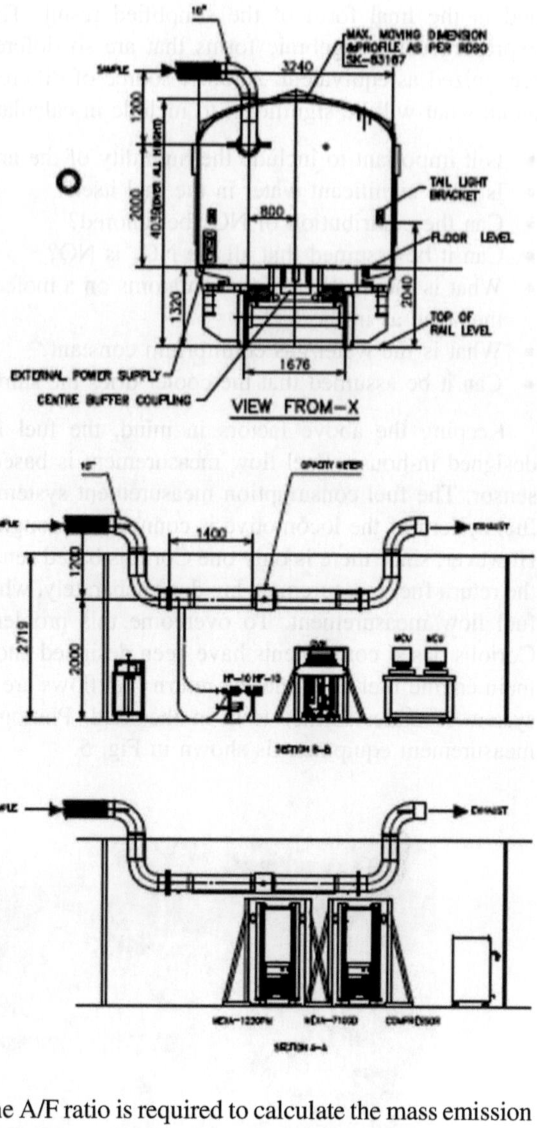

measured in terms of ppm and the A/F ratio is required to calculate the mass emission values. There are an uncountable number of different equations for calculating the air to fuel ratio of engine's combustion process from its exhaust emissions. There are 72 such equations collected and supported by the MEXA 9000 of Horiba Make (installed gaseous emissions measurement equipment in the ETC). Authors such as D'Alleva, Spindt, Brettschneider, Lange, Simons, Stivender and others have published papers that are often referenced as sources for these equations.

Most of the differences between equations are a matter of algebra. Since many expressions arise during the derivation of the A/F equation, there is much room for creativity in the selection of the algebraic steps taken in the simplification process and in the final form of the simplified result. The very same equation can be expressed using algebraic forms that are so different that they can no longer be recognized as equivalent. Another source of differences are the basic assumptions about what will be significant to include in calculation. For example:

- Is it important to include the humidity of the ambient air?
- Is there significant water in the fuel itself?
- Can the contribution of NO_x be ignored?
- Can it be assumed that all the NO_x is NO?
- What is the number of carbon atoms on a molecule of HC in the exhaust? Is it the same as in the fuel?
- What is the water/gas equilibrium constant?
- Can it be assumed that the cooler dries the sample completely?

Keeping the above factors in mind, the fuel flow measurement system was designed in-house. Fuel flow measurement is based on a Coriolis principle based sensor. The fuel consumption measurement system is portable. The fuel from the fuel system of the locomotive is connected through this fuel measurement system. However, since there is only one Coriolis based sensor, main fuel measurement and the return fuel measurement are done separately, which may cause some error in the fuel flow measurement. To overcome this problem a fuel flow meter with two Coriolis based components have been designed and successfully tested. In this the main engine fuel flow and the return fuel flows are measured simultaneously. This system is under further tests in the field. Photograph of the existing fuel flow measurement equipment is shown in Fig. 5.

Fig. 5 Existing fuel flow measurement system

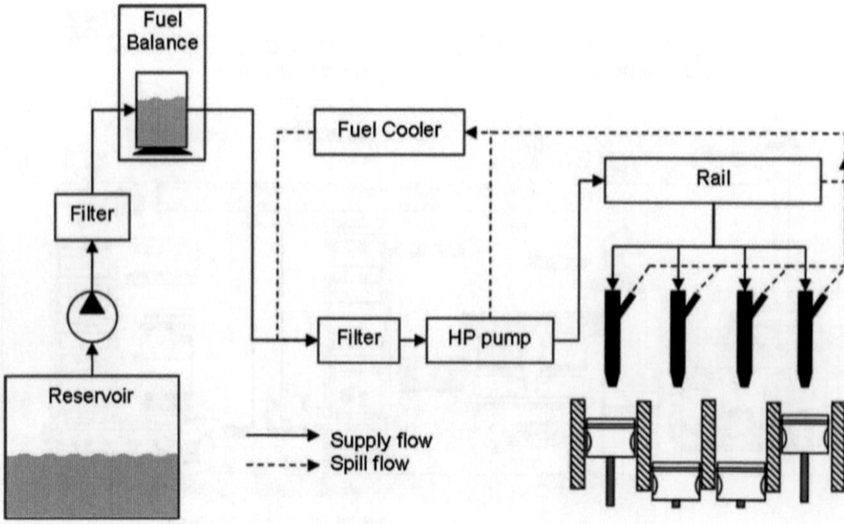

Fig. 6 Schematic of the fuel flow measurement system for locomotive engine [8]

Schematic of the fuel flow measurement system for a locomotive engine is shown in Fig. 6. Supply flow and spill flow are measured separately. Spill flow is deducted from the supply flow to arrive at the engine fuel consumption. The fuel flow meter is provided with the following components:

- Shut off valve
- Coarse filter
- Fine filter
- Fuel pump
- Consumption sensor
- Venting valve
- Pressure regulator
- Pressure gauge
- Fuel cooler

The fuel flows from the supply (i.e. day tank) via shut-off valve into the measurement system. A course filter with 200 μ mesh followed by a fine filter with a 60 μ mesh filter on the pump inlet side. The fuel pump circulates the fuel by passing the back-pressure regulator. The return fuel is fed back into the day tank. When the engine consumes fuel, the required volume flows via Coriolis consumption sensor to the engine. The fuel is kept circulating on the engine side by the engine's pre-pump. The fuel pressure in the feed and return lines is adjusted by pressure regulator.

Figure 7 gives the final layout of the emission test car. Exhaust is collected from the exhaust chimney of the locomotive engine and transported to the ETC by using a connecting pipe.

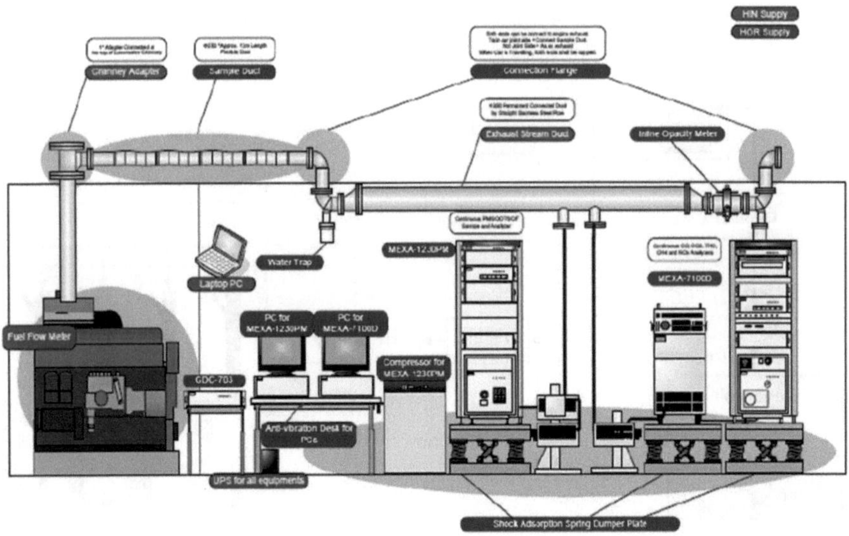

Fig. 7 Final layout of the emission test car [9]

The connecting pipe outputs into the fixed exhaust pipe of the ETC, the fixed pipe of the ETC has a suction pump at the end to control the flow of the exhaust. Fixed pipe of the ETC has sampling tubes attached to it which provide the samples to the online realtime PM system and the Gaseous measurement system. Rest of the exhaust gas is exhausted out of the suction pump (Fig. 7).

2.4 Design for the Shock and Vibrations Encountered During Run

The environmental conditions for the railway vehicles are harsh, they have to face extreme humidity, dust conditions, heat and cold weathers. The temperatures range from −10 to 50 °C and the vibrations and shocks to 3G in all the three directions. Thus, it was necessary to design a shock and vibration system which is capable of withstanding the running environment [10]. A wire vibration isolator assembled with a rack cabinet is shown in Fig. 8.

Wire rope Isolators are comprised of stainless steel stranded cable, threaded through aluminium alloy retaining bars, crimped and mounted for effective vibration isolation (Fig. 9). With their corrosion resistant, all metal construction, the wire rope isolators are environmentally stable, high-performance and shock vibration isolators that are unaffected by temperature extremes, chemicals, oils ozone and abrasives. Wire rope isolators were used to isolate the shocks and vibrations. Air conditioning of the equipments room was used to control the weather conditions.

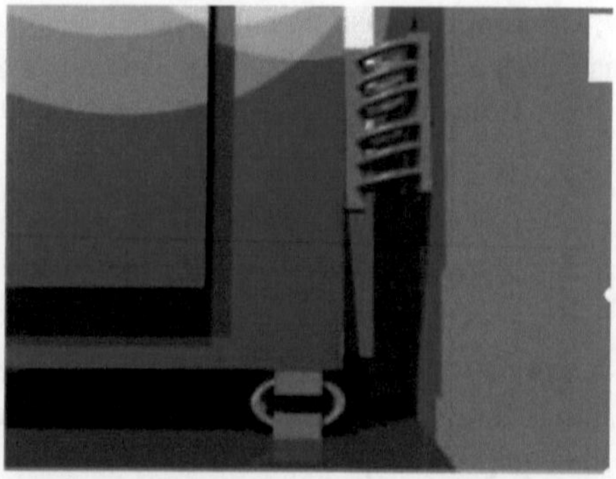

Fig. 8 Wire rope isolators used on the ETC for various equipment

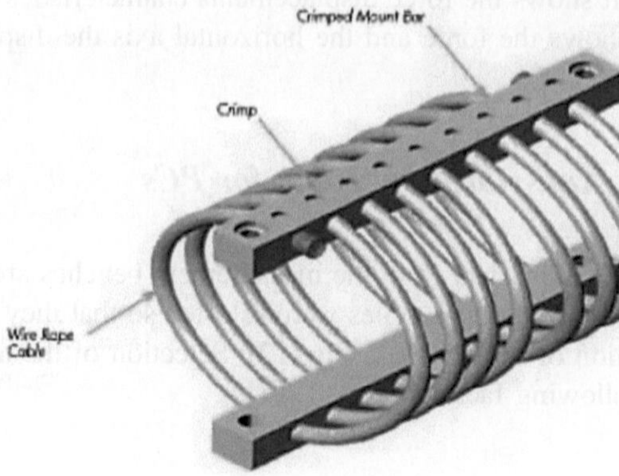

Fig. 9 Wire vibration isolator

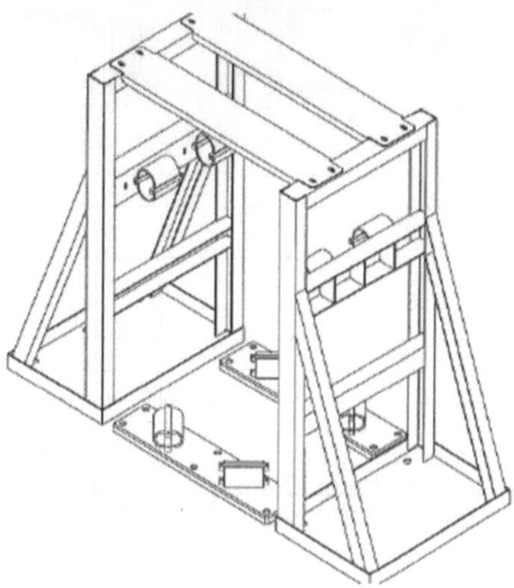

Fig. 10 Isolating frames for emission measurement benches

Emission measurement benches were mounted in isolating frames with wire isolators as shown in Fig. 10. This allowed the shocks and vibrations to be effectively isolated from the emission measurement equipment.

2.5 Anti-vibration Calculations

A sample of the Anti-vibration calculations are shown in Fig. 11. This has taken into account the total weight of the equipment, vibration levels to be encountered by the equipment, speed and direction of the vibrations, average impact force. These inputs are used to calculate the number of isolators and the load weight per isolator. Isolators characteristics and selection are done based on the above inputs. Graph on the left shows the force displacements characteristics of different isolators. Vertical axis shows the force and the horizontal axis the displacement of the isolator.

2.6 Anti-vibration Table for PCs

Measurement data from the measurement benches are displayed on the PCs in real time. Anti-vibration tables were selected so that they can withstand the shocks and vibration of the coaches (Fig. 12). Selection of the anti-vibration table considered the following factors:

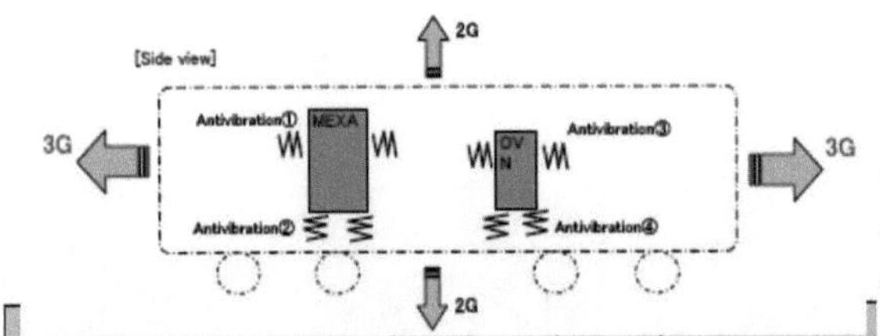

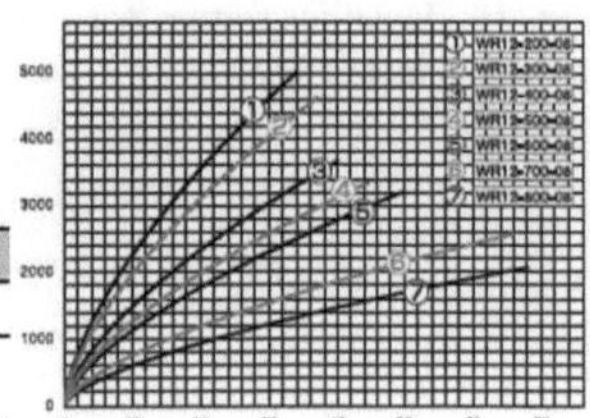

全荷重 Total Weight	Wt	3430	[N]	Wt = 350 [kg] * 9.8
防振器の使用数 The number of Dumper.	n	2	[PC]	
防振器一個あたりの支持荷重 Load weight per dumper	W	1715	[N]	W = Wt/n = 3430 / 2
設置方向 Direction		圧縮		
落下高さ Height	h	50	[mm]	
加速度 Accurate	Gt	3	[G]	
速度 Speed	V	0.9899	[m/sec]	V = √(2gh)
最小たわみ Minimum flexible volume	D	50	[mm]	D = (V² / g (Gt−1))*1000
平均衝撃荷重 Average Impact Force	Favg	1715	[N]	Favg = (WV² / 2gD)*1000
平均たわみ Average flexible volume	davg	25	[mm]	davg = D / 2

Fig. 11 Anti-vibration calculations

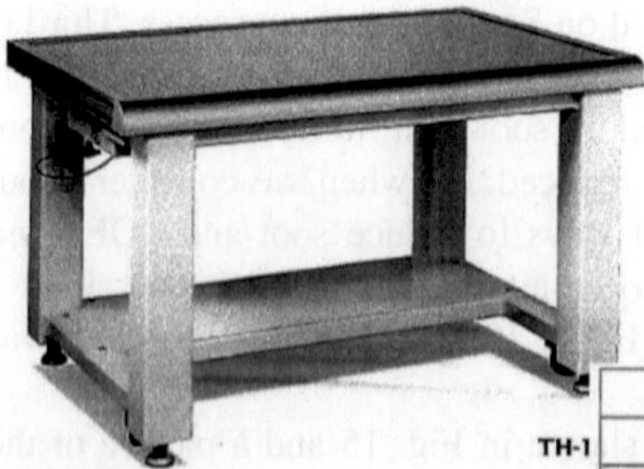

SPEC	TH-600	TH-1130	TU-600	TU-1130
Natural Frequency	(Vertical) 1.5~2.5Hz			
Vibration Isolating Method	Diaphragm type air spring			
Leveling Retention Method	A:Automatic leveling (air supply pressure 0.25~0.5MPa)			
Surface Table	High rigid steel honeycomb surface		Stainless steel finish	
Surface Table size (mm)	600×700	1130×700	600×700	1130×700
Max.Loading Capacity (kg)	2[illegible]0			
Dimension (mm)	1200W×8[illegible]0D×750H			
Weight (kg)	125	145	112	120
Accessories included	Piping tube (φ6mm×5m), One-touch control joint PT1/4 male screw			
Option	Drop-preventive bottom structure・Concentrated exhaust outlet			

Fig. 12 Anti-vibration table for the PCs

- Natural frequency of vibration
- Vibration isolating method
- Levelling retention method
- Surface table finish and size
- Maximum loading capacity

2.7 *Online PM Measurement System [10, 11]*

Particulate Matter from diesel engine emissions consists of carbon, hydrocarbon, SO_4 and H_2O. Figure 13 shows that PM consists of mostly 3 components. First is soot, which is carbonaceous solid. In engine combustion process, carbon particle is formed in fuel-rich region. Second is SOF, soluble organic fraction, which made of heavier hydrocarbon. Heavier hydrocarbon is derived from unburned fuel and oil.

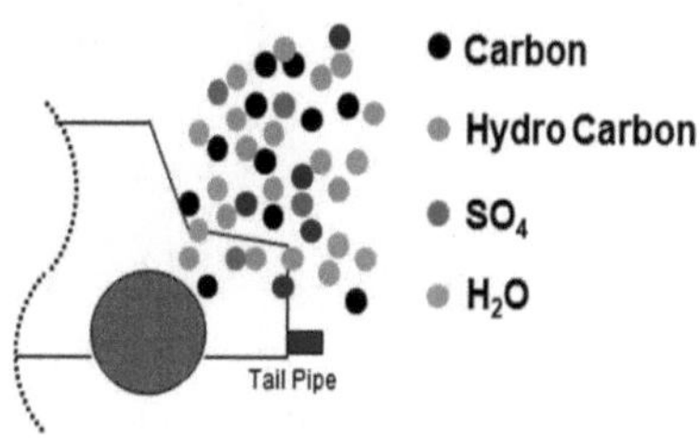

Fig. 13 Constituents of particulate matter (PM)

Heavier hydrocarbon is condensed or absorbed on Soot in dilution process. Third is sulfates. Sulfur in fuel or lube oil is oxidized to SO_3. In dilution process, SO_3 is condensed, combined with H_2O and absorbed on soot. But Sulfates decrease more and more, because sulfur in fuel and oils is reduced. So when we consider about reduction of PM, we have to think different ways to reduce soot and SOF. And therefore it is important for diesel industry to monitor soot and SOF separately.

For measurement of particulate matter, MEXA 1230 PM of Horiba Make was used (Fig. 14).

Working concept for MEXA-1230PM is shown in Fig. 15 and a picture of the bench is shown in Fig. 14. Sample is diluted by an ejector pump, by using the coefficient (Fig. 16), soot mass concentration can be calculated from DC sensor output. SOF is measured by the difference of the THC concentration by using double FIDs. By using the coefficient (Fig. 17), SOF mass concentration can be calculated from the difference of THC concentration.

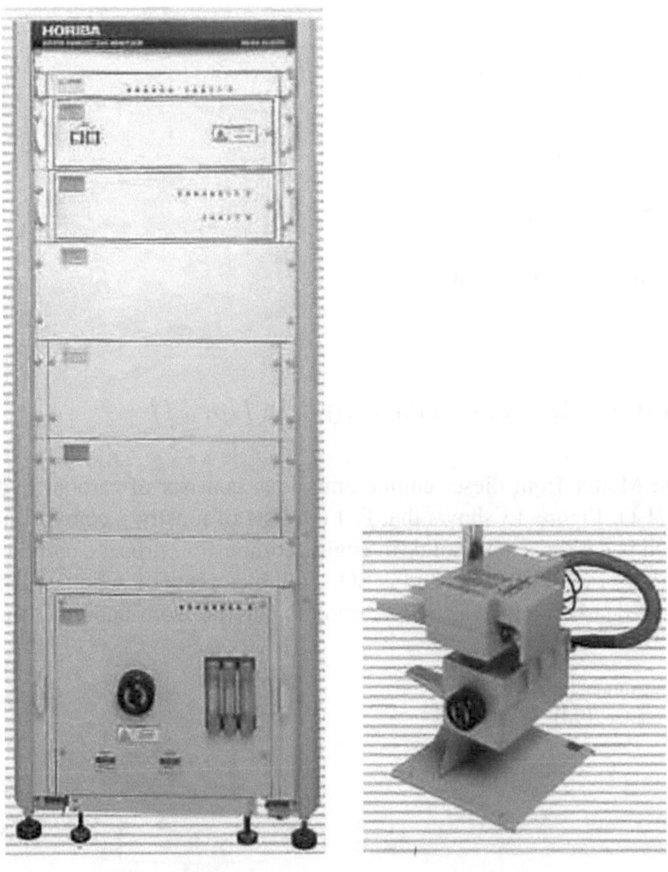

Fig. 14 MEXA 1230 PM used to measure the online real time PM, soot and SOF

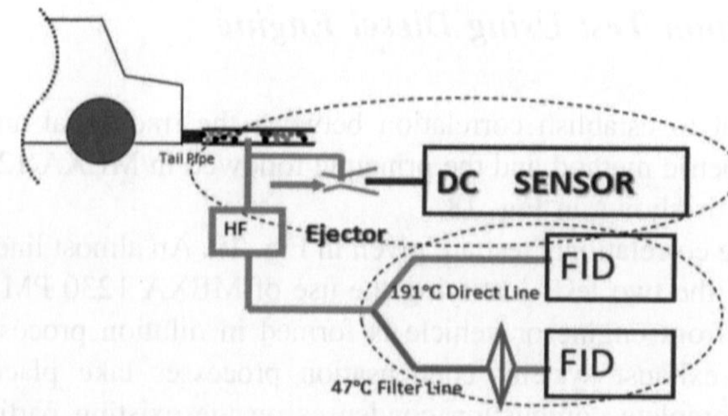

Fig. 15 Working concept for Horiba Make MEXA 1230 PM

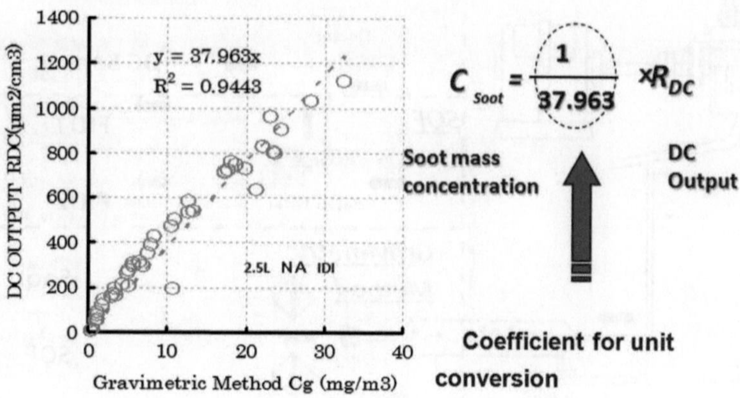

Fig. 16 Coefficient calculation for soot concentration

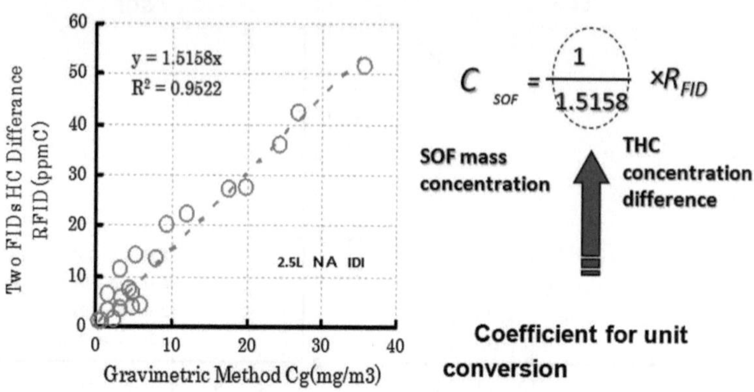

Fig. 17 Coefficient calculation for SOF concentration

2.8 Correlation Test Using Diesel Engine

It was important to establish correlation between the traditional and universally accepted gravimetric method and the principle followed in MEXA 1230 PM. A set up for the same is shown in Fig. 18.

Results of the co-relational test are given in Fig. 19. An almost linear correlation is seen between the two tests justifying the use of MEXA 1230 PM (Fig. 19).

PM emitted from engine or vehicle is formed in dilution process. During the cooling in the exhaust system, condensation processes take place. Vapour of products of incomplete combustion condenses on the existing particles and new

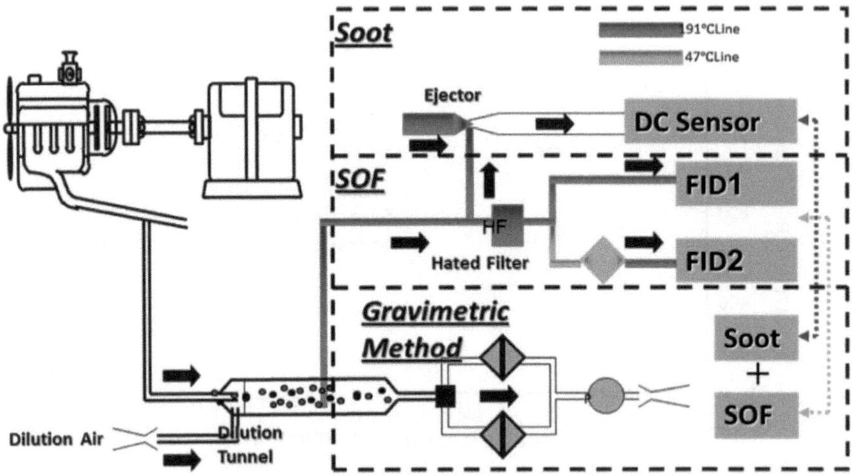

Fig. 18 Engine set-up for the co-relational test

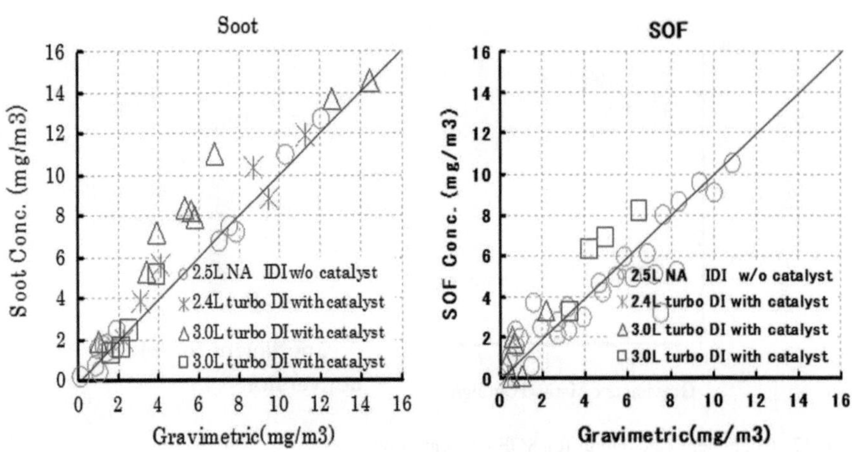

Fig. 19 Correlational tests for the two method of measuring PM

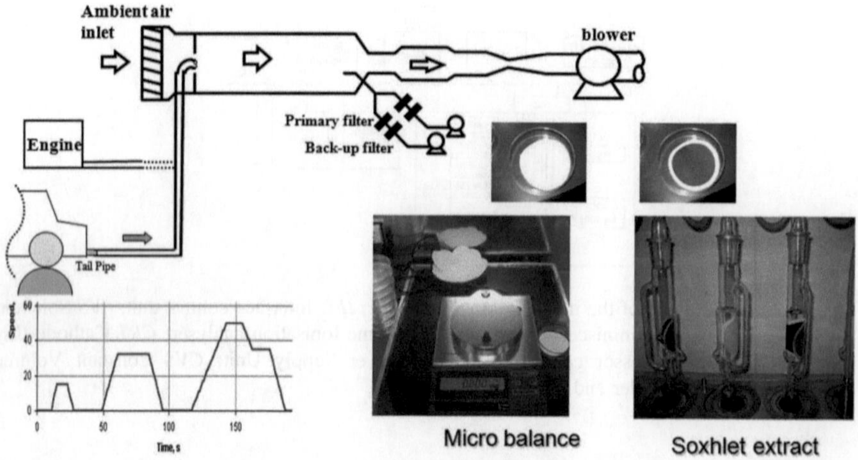

Fig. 20 Gravimetric method of measuring particulate matter, soot and SOF

volatile droplets can be formed by nucleation. The agglomerates may therefore carry a condensed layer of hydrocarbons and/or sulphur compounds on their surfaces.

The present, PM measurement is carried out by gravimetric method. First, the exhaust gas from the engine or tail pipe is introduced to the partial dilution tunnel. Then the gas is diluted and temperature is cooled down. From this point, the PM in the diluted exhaust is introduced to and deposited on the filter. The weight of the filters is measured by a micro balance (Fig. 20). But before measurement, the filters have to be settled in the chamber at least 8 h. The chamber is kept at continuous temperature and humidity, to separate soot and SOF, soxhlet extract is used, it takes at least 4 h, and before and after the treatment, the weight is measured, so it takes about 3 days to carry out the above procedure.

There are many demerits of using the gravimetric method. First and foremost is that it takes long time to get results, especially soot and SOF separation. Second demerit is that it is a batch test only and not real time. Third demerit is that it is difficult to analyse low emission samples. Besides, it was not possible to install the gravimetric equipment because the balance requires complete stability, which is not possible to provide in a mobile emission test car. It was decided to have an online real time soot, SOF and PM measurement system and MEXA 1230PM of Horiba Make was selected after due consideration. The advantages of this equipment are that it can measure mass and the rate of component in low PM emission, measure continuously soot and SOF separately and measure low PM after DPF, oxidization catalyst.

2.9 Gaseous Pollutants Measurement

MEXA 7000 a standard gaseous emissions measurement system of Horiba Make was used for measurement of the and general layout of the modules is shown in Fig. 21. Picture of the analyser bench is shown in Fig. 22 [12].

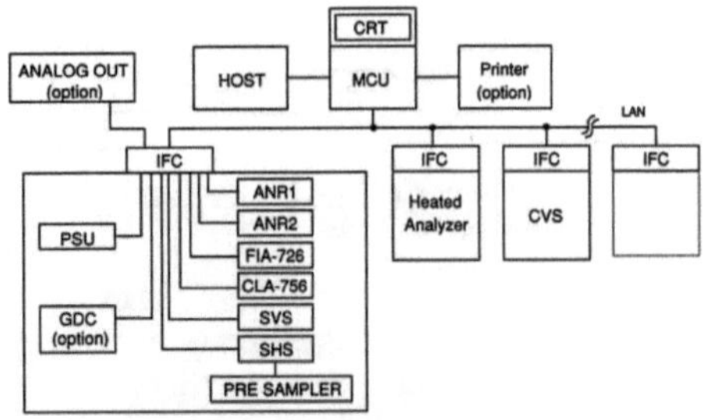

Fig. 21 General layout of the modules. *ANR* Analyser; *IFC* Interface control unit; *SVS* Solenoid valve unit; *CLA* Chemiluniniscence analyser; *FIA* Flame Ionisation analyser; *CRT* Cathode Ray Tube; *MCU* Microprocessor control unit; *PSU* Power Supply Unit; *CVS* Constant Volume sampler; *GDC* Gas Divider and conditioning unit

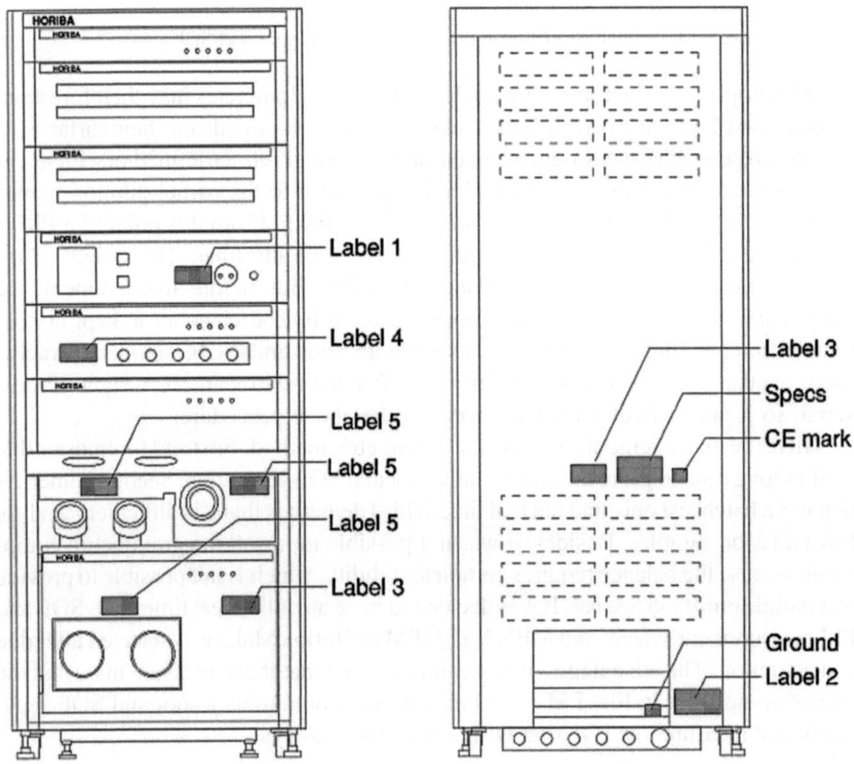

Fig. 22 MEXA 7000 gas analyser unit of Horiba make

Table 9 List of operation and measurement gases used for MEXA 7000

Gas type	Gas concentration	Use flow	Use flow/ANA1pcs	Pressure	
Operation gas (MEXA-7100D & 1230PM)					
O_2	O_2 99.99% (not more than 3 ppmC for THC/0.1 ppm for No_x)	0.4 L/min	0.4 Lmin (CLA)	100 ± 10 kPa	
40% $H_{2/He}$(FUEL)	H_2 (39.0–41 vol.%)/He base (not more than 0.3 ppmC for THC)	0.45 L/min	0.15 L/min, 0.3 L/min	100 ± 10 kPa	
AIR	O_2 200–21.0 vol.% (not more than 0.1 ppmC for THU/10 ppm for CO_2/1 ppm for $CO.NO_x$)	0.5 L/min	0.2 L/min, 0.3 L/min	100 ± 10 kPa	
Calibration gas (MEXA-7100D & 1230PM)					
Gas type	Gas concentration	Analyzer	Use flow	Ranges	Pressure
CO(L)	(4500 ppm) Cal range scale × 85–95%/N_2 base	AIA-721A	3 L/min	0···0.50–500 ppm	100 ± 10 kPa
CO(H)	(10.8%) Cal range scale × 85–95%/N_2 base	AIA-722	2 L/min	0···.0.5–12 vol. %	100 ± 10 kPa
CO_2	(18%) Cal range scale × 85–95%/N_2 base	AIA-752	2 L/min	0···.0.5–20 vol. %	100 ± 10 kPa
T.HC(C_3H_8)	(450 ppmC) Cal range scale × 85–95%: air base	FIA-725 A	1 L/min	0–10–500 ppmC	100 ± 10 kPa
T.HC(C_3H_8)	(4500 ppmc) Cal Range Scale × 85–95%: air base	FIA-725A	1 L/min	0.0–1000–50000 ppmC	100 ± 10 kPa
CH_4	(45 ppm) Cal range scale × 85–95%/N_2 base	GFA-720	1 L/min	0–5–50 ppm	100 ± 10 kPa
CH_4	(2250 ppm) Cal range scale × 85–95%/N_2 base	GFA-720/FIA-721HA	1 L/min	0–50–2500 ppmC	100 ± 10 kPa
CH_4	(22,500 ppm) Cal range scale × 85–95%/N_2 base	FIA- 721 HA	1 L/min	0–5000–25000 ppmC	100 ± 10 kPa
NO	(450 ppm) Cal range scale × 85–95%/N_2 base	CLA-755A	1 L/min	0–10–500 ppm	100 ± 10 kPa
NO	(9000 ppm) Cal range scale × 85–95%/N_2 base	CLA-755A	1 L/min	0–1000–10000 ppm	100 ± 10 kPa

Fig. 23 Gas room of the ETC

List of various utility equipments fitted on the ETC is shown in Table 9. Having a stable voltage regulator on the coach was one of the main issues.

A picture of the gas room designed for the ETC is shown in Fig. 23.

3 Test Results

It was necessary to co-relate the emission measurement data from the Engine Test Beds Mass Emission Measurement System (MEMS) to that obtained from the ETC. Pictures of the locomotive emission measurement are shown in Fig. 24 [13].

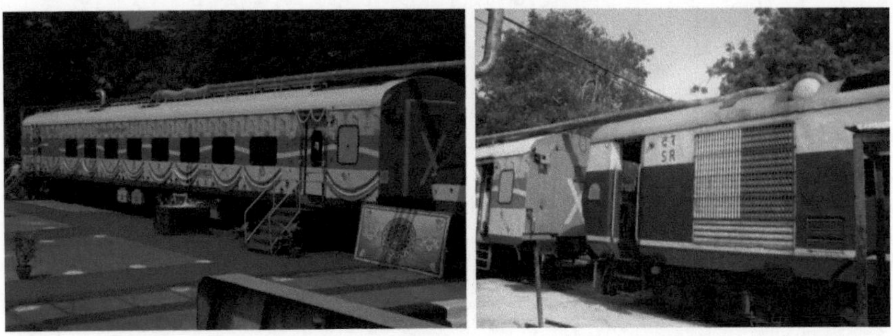

Fig. 24 Measurement of emission by ETC from a diesel multiple unit [14]

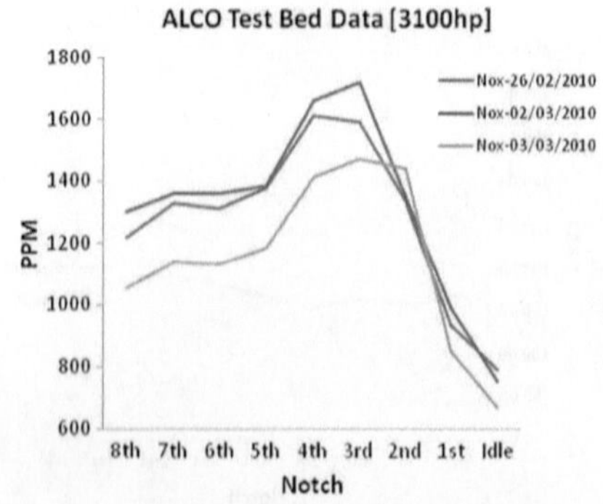

(a) Alco test bed data NOx Trends

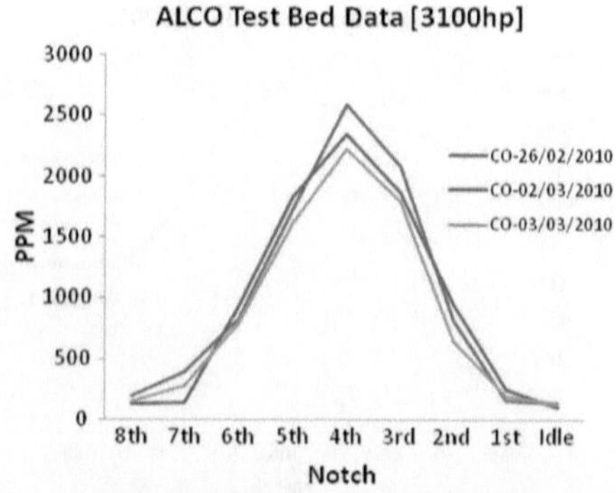

(b) Alco test bed data CO Trends

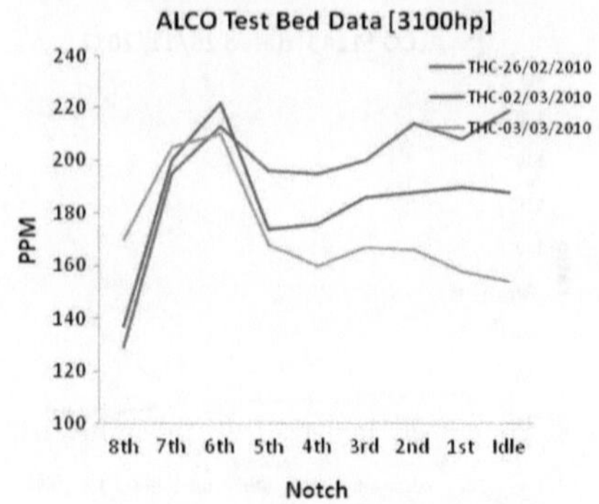

(c) Alco test bed data THC Trends

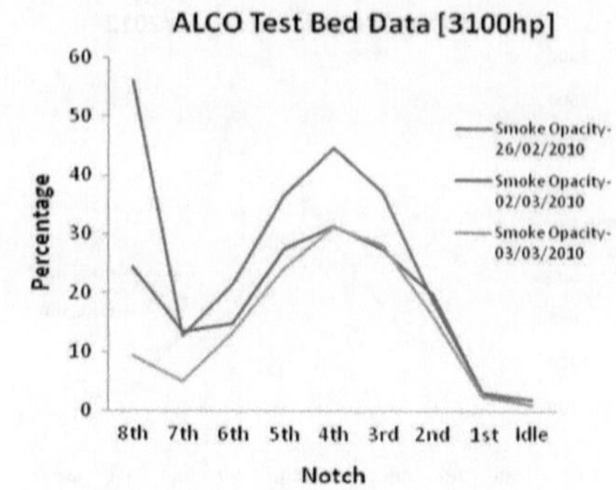

(d) Alco test bed data Smoke Opacity Trends

Fig. 25 Emission test results from the engine test bed MEMS [14]

Figure 25 gives the measurement results from the Engine Test Bed MEMS. Figure 26 gives the measurement results for a diesel locomotive of the ETC Mass Emission Measurement System (MEMS). It was expected to have differences in the values of the measurement results due to different configurations of the test bed engine and the locomotive engine. However the general trends in NO_x, THC and CO as well as the values appear to be similar. However, there are some difference in the opacity levels mainly due to different configurations and environmental conditions between the test bed and the diesel locomotive.

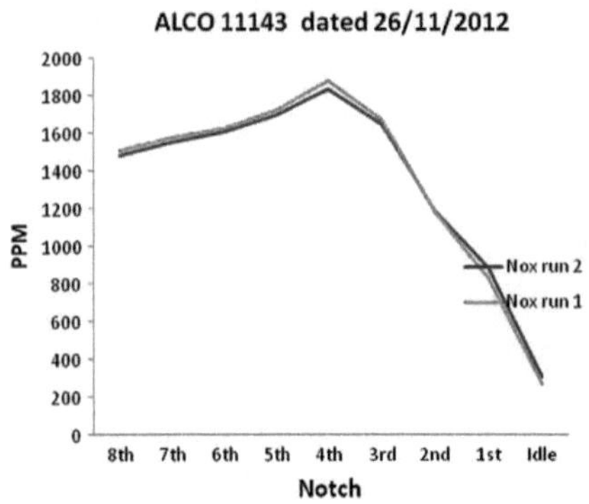

NOx Trend

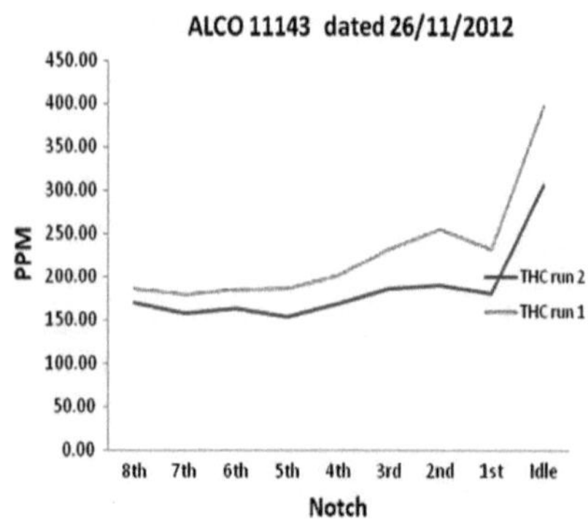

THC Trend

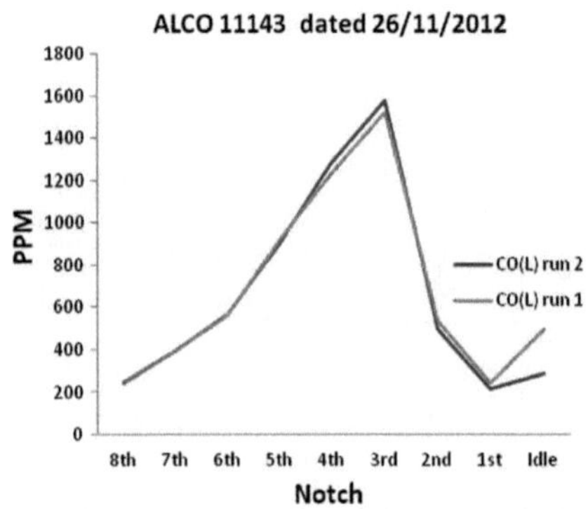

CO (L) Trend

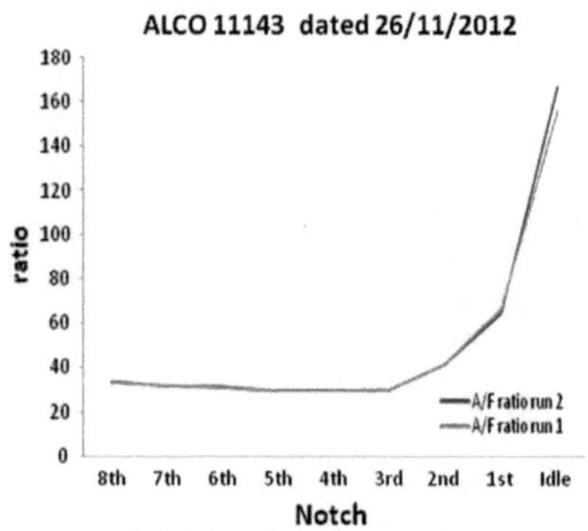

Air/Fuel Ratio Trend

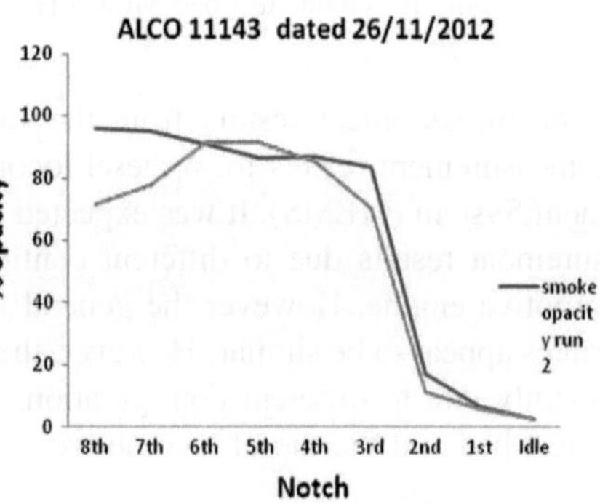

Smoke opacity Trend

Fig. 26 Emission test results from the ETC MEMS

4 Conclusions

A mobile emission test car for measuring regulated emissions from diesel locomotives of Indian Railways has been designed and developed by RDSO and ICF under the Ministry of Railways (MoR). It is the first kind of mobile emission test car in the world and a patent has been filed for the same. A distinguishing feature of the ETC is the use of wire isolators and frames to prevent the scientific grade equipments from shocks and vibrations encountered during train run. Another significant attribute of the ETC is the use of on-line real time PM measuring system. This can be used for measuring PM during engine transients when the smoke levels become higher. The ETC layout has been designed by considering the volume and the weight of the equipments and the Air conditioning capacity has been determined based on the heat load of the equipments as well as the living quarters. ETC is being presently used to measure emissions from diesel locomotives of Indian Railways. This data shall be used to arrive at the emission standards for the diesel locomotives of IR and also to monitor the emission compliant developed locomotives of IR.

References

1. Railways year book 2014-15
2. National Green Tribunal order on diesel locomotives of IR
3. US EPA locomotive pollution regulation
4. European locomotive pollution regulation
5. ISO 8178—Pollution measurement in steady state from non-road machinery and vehicles
6. www.diesel.net
7. ICF drawing of the ETC
8. Fuel measurement system manual
9. Horiba drawing for approval
10. Horiba document submitted for approval to RDSO
11. MEXA 1230 PM manual of Horiba
12. Presentation made by Horiba on PM 1230
13. MEXA 7000 manual of Horiba
14. RDSO internal report on ETC, 2010

MIX
Papier aus verantwortungsvollen Quellen
Paper from responsible sources
FSC® C105338

If you have any concerns about our products,
you can contact us on
ProductSafety@springernature.com

In case Publisher is established outside the EU,
the EU authorized representative is:
Springer Nature Customer Service Center GmbH
Europaplatz 3, 69115 Heidelberg, Germany

Printed by Libri Plureos GmbH
in Hamburg, Germany